Lecture Notes in Computer Science 13819

More information about this series at https://link.springer.com/bookseries/558

Yi Chang · Xiaofei Zhu (Eds.)

Information Retrieval

28th China Conference, CCIR 2022
Chongqing, China, September 16–18, 2022
Revised Selected Papers

Editors
Yi Chang
Dingxin Building C403
Jilin University
Changchun, China

Xiaofei Zhu
College of Computer Science
and Engineering
Chongqing University of Technology
Chongqing, China

ISSN 0302-9743 ISSN 1611-3349 (electronic)
Lecture Notes in Computer Science
ISBN 978-3-031-24754-5 ISBN 978-3-031-24755-2 (eBook)
https://doi.org/10.1007/978-3-031-24755-2

This Springer imprint is published by the registered company Springer Nature Switzerland AG
The registered company address is: Gewerbestrasse 11, 6330 Cham, Switzerland

Preface

The 2022 China Conference on Information Retrieval (CCIR 2022), sponsored by the Chinese Information Processing Society of China (CIPS), organized by the Information Retrieval Committee of the Chinese Information Processing Society of China (CIPS) and Chongqing University of Technology, co-organized by Chongqing Artificial Intelligence Association, was the 28th installment of the conference series. The conference was held in Chongqing on September 16–18, 2022.

Many famous experts and scholars from China and overseas were invited to the conference to deliver special reports. The conference also scheduled paper presentations, workshops on hot research issues, a young scholar forum, a new committee forum, an industry forum, evaluation activities, and poster exchanges, and Chinese authors of international journals and conferences (such as TOIS, SIGIR, WWW, CIKM and ACL) communicated papers. During the conference, there was a meeting of the Retrieval Committee of the Chinese Information Processing Society of China (CIPS).

A total of 109 papers were submitted for this conference. In order to ensure the quality of the review, the Procedure Committee organized 84 experts and scholars in the field of information retrieval to participate in the review, so as to ensure that each paper received fair and reasonable high-level feedback from at least three experts in the field. Finally, a total of 63 papers were accepted for presentation at the conference, an acceptance rate of 57.8%. The accepted Chinese-language papers will be recommended for publication in relevant academic journals, and the eight accepted English-language papers are compiled and published in this Springer LNCS series volume.

This year, the conference continued the basic format of last year, continuing to call for papers in English, and we collaborated with Springer to publish the proceedings. The conference invited the authors of papers recently published in the field of information retrieval at top international conferences to introduce and exchange related work. The paper report structure of the conference is a single-venue system, which ensures the efficiency and quality of the conference communication in the form of high-level paper conference reports + poster exchange of all accepted papers. Social media was used for online communication, and selected parts of the conference were reported on-site.

The Program Committee would like to thank the diamond sponsors iFlyTek, Beijing Kuaishou Technology, the gold sponsor Chongqing Shouheng Software, the silver sponsor Baidu, and the bronze sponsor Beijing Bayou Technology for their contributions to this conference. Special thanks to Gao Xinbo, Yin Dawei, ChengXiang Zhai and Cao Chao for their excellent guest reports. Thanks to Liu Kang and Zhang Weinan for organizing the wonderful forum and workshop for young scholars. Thanks to Wang Pengfei for publishing the papers. Thanks to Rosimbo and Yixing Fan for the publicity work of the conference. Thanks to Ren Zhaochun and Luo Cheng for conference sponsorship-related work. Thanks to Zhang Jianxun, Huang Xianying, Xiao Zhaohui, Huang Lifeng and other teachers from Chongqing University of Technology for undertaking the complicated work of organizing the conference. Thanks to Louie, Ma Weizhi,

Iflytek, China Mobile Research Institute, China Mobile Technology Capability Evaluation Center, National University of Singapore, University of Science and Technology of China, Sichuan University, Singapore 6Estates Company, Beijing Maijia Intelligent Technology efforts to organize the "Iflytek-National Information Retrieval Challenge Cup". Sincere thanks to all the contributors, process committee members, delegates and all the volunteers who worked for CCIR 2022.

We would also like to thank the editorial team of Springer, the Journal of Computers, Computer Research and Development, Pattern Recognition and Artificial Intelligence, the Journal of Chinese Information, Computer Science and Exploration, the Journal of South China University of Technology (Natural Science Edition), the Journal of Shandong University (Science Edition), the Journal of Shanxi University (Natural Science Edition), the Journal of Guangxi Normal University, the Journal of Chongqing University of Technology and other journals for their efforts and hard work on the publication of the Chinese-language and English-language papers of this conference.

The China Conference on Information Retrieval (CCIR) is the most important meeting in China in the field of information retrieval. The meeting, using "create the new engine of the digital economy" as the main theme, consisted of a series of academic activities and focused on a new generation of scientific research on information retrieval technologies. It also provided a wide range of communication platforms for the dissemination of the latest academic research results in the information retrieval area. Academic study of the information retrieval field aims to meet the demand of humans on the Internet to obtain information quickly and accurately: research results will tend to support the national strategy, promote the development of the Internet and artificial intelligence, provide strong support for the digital economy development of our country, improve the production efficiency of the whole society, and make a significant impact on various fields of social life.

We believe this session was a complete success.

September 2022

Yi Chang
Xiaofei Zhu

Organization

General Chairs

Jirong Wen	Renmin University of China, China
Guoxi Wang	Chongqing University of Posts and Telecommunications, China
Xiaokang Liu	Chongqing University of Technology, China

Program Committee Chairs

Yi Chang	Jilin University, China
Xiaofei Zhu	Chongqing University of Technology, China

Proceedings Chairs

Junjie Wen	Chongqing University, China
Wu Yang	Chongqing University of Technology, China
Zhigang Zhang	Chongqing University of Technology, China

Publicity Chair

Pengfei Wang	Beijing University of Posts and Telecommunications, China

Webmaster

Jianxun Zhang	Chongqing University of Technology, China
Xianying Huang	Chongqing University of Technology, China

Youth Forum Chairs

Kang Liu	Institute of Automation, Chinese Academy of Sciences, China
Weinan Zhang	Harbin Industrial University, China

CCIR Cup Chairs

Weidong Liu — China Mobile Communication Research Institute, China
Weizhi Ma — Tsinghua University, China

Sponsorship Chairs

Zhaochun Ren — Shandong University, China
Cheng Luo — MegaTech.AI, China

Treasurers

Chaohui Xiao — Chongqing University of Technology, China
Lifeng Huang — Chongqing University of Technology, China

Awards Chair

Shoubin Dong — South China University of Science and Engineering, China

Program Committee

Fei Cai — National University of Defense Technology, China
Jiawei Chen — University of Science and Technology of China, China
Xiaoliang Chen — Xihua University, China
Yubo Chen — Institute of Automation, Chinese Academy of Sciences, China
Zhumin Chen — Shandong University, China
Zhicheng Dou — Renmin University of China, China
Yajun Du — Xihua University, China
Yixing Fan — Institute of Computing Technology, Chinese Academy of Sciences, China
Shengxiang Gao — Kunming University of Science and Technology, China
Zhongyuan Han — Foshan University, China
Yanbin Hao — University of Science and Technology of China, China
Ben He — University of Chinese Academy of Sciences, China
Xiangnan He — University of Science and Technology of China, China

Yu Hong	Soochow University, China
Zhuoren Jiang	Zhejiang University, China
Ting Jin	Hainan University, China
Yanyan Lan	Tsinghua University, China
Chenliang Li	Wuhan University, China
Lishuang Li	Dalian University of Technology, China
Ru Li	Shanxi University, China
Shangsong Liang	Sun Yat-sen University, China
Xiangwen Liao	Fuzhou University, China
Hongfei Lin	Dalian University of Technology, China
Yuan Lin	Dalian University of Technology, China
Chang Liu	Peking University, China
Peiyu Liu	Shandong Normal University, China
Cheng Luo	Tsinghua University, China
Zhunchen Luo	PLA Academy of Military Science, China
Jianming Lv	South China University of Technology, China
Ziyu Lyu	Shenzhen Institute of Advanced Technology, Chinese Academy of Sciences, China
Weizhi Ma	Tsinghua University, China
Jiaxin Mao	Tsinghua University, China
Xianling Mao	Beijing Institute of Technology, China
Liqiang Nie	Shandong University, China
Liang Pang	Institute of Computing Technology, Chinese Academy of Sciences, China
Zhaochun Ren	Shandong University, China
Tong Ruan	East China University of Science and Technology, China
Ting Bai	Beijing University of Posts and Telecommunications, China
Huawei Shen	Institute of Computing Technology, Chinese Academy of Sciences, China
Dawei Song	Beijing Institute of Technology, China
Ruihua Song	Renmin University of China, China
Xuemeng Song	Shandong University, China
Hongye Tan	Shanxi University, China
Songbo Tan	Beijing Hengchang Litong Investment Management, China
Liang Tang	Information Engineering University, China
Hongjun Wang	TRS Information Technology, China
Pengfei Wang	Beijing Institute of Technology, China
Suge Wang	Shanxi University, China
Ting Wang	National University of Defense Technology, China

Zhiqiang Wang	Shanxi University, China
Zhongyuan Wang	Meituan, China
Dan Wu	Wuhan University, China
Le Wu	Hefei University of Technology, China
Yueyue Wu	Tsinghua University, China
Jun Xu	Renmin University of China, China
Tong Xu	University of Science and Technology of China, China
Weiran Xu	Beijing Institute of Technology, China
Hongbo Xu	Institute of Computing Technology, Chinese Academy of Sciences, China
Kan Xu	Dalian University of Technology, China
Hongfei Yan	Peking University, China
Xiaohui Yan	Huawei, China
Muyun Yang	Harbin Institute of Technology, China
Zhihao Yang	Dalian University of Technology, China
Dongyu Zhang	Dalian University of Technology, China
Hu Zhang	Shanxi University, China
Min Zhang	Tsinghua University, China
Peng Zhang	Tianjin University, China
Qi Zhang	Fudan University, China
Ruqing Zhang	Institute of Computing Technology, Chinese Academy of Sciences, China
Weinan Zhang	Harbin Institute of Technology, China
Ying Zhang	Nankai University, China
Yu Zhang	Harbin Institute of Technology, China
Chengzhi Zhang	Nanjing University of Science and Technology, China
Xin Zhao	Renmin University of China, China
Jianxing Zheng	Shanxi University, China
Qingqing Zhou	Nanjing Normal University, China
Jianke Zhu	Zhejiang University, China
Xiaofei Zhu	Chongqing University of Technology, China
Zhenfang Zhu	Shandong JiaoTong University, China
Jiali Zuo	Jiangxi Normal University, China

Contents

A Position-Aware Word-Level and Clause-Level Attention Network for Emotion Cause Recognition

Yufeng Diao[1(✉)], Liang Yang[2], Xiaochao Fan[3], and Hongfei Lin[2]

[1] Inner Mongolia Minzu University, Tongliao 028043, China
542275403@qq.com
[2] Dalian University of Technology, Dalian 116024, China
[3] Xinjiang Normal University, Urumq 830054, China

Abstract. Emotion cause recognition is a vital task in natural language processing (NLP), which aims to identify the reason of emotion expressed in text. Both industry and academia have realized the importance of the relationship between emotion word and context. However, most existing methods usually ignore the fact that the position information is also crucial for detecting the emotion cause. When an emotion word occurs in a clause, its neighboring words and clauses should be given more attractive than others with a long distance. In this paper, we propose a novel framework Position-aware Word-level and Clause-level Attention (PWCA) Network based on bidirectional GRU. PWCA not only concentrates on the position information of emotion word, but also builds the relation between emotion clause and candidate clause by leveraging word-level and clause-level attention mechanism. The experimental results show that our model obviously outperforms other state-of-the-art methods. Through the visualization of attention over words, we validate our observation mentioned above.

Keywords: Emotion cause recognition · Position-aware · Word-level · Clause-level · Attention · Bidirectional GRU

1 Introduction

Emotion cause recognition is one of the significant tasks in natural language processing (NLP), which focuses on the analysis of emotion causes in texts, especially in online internet platforms. It aims to detect the reason from a certain emotion expression in a text, and has received plenty of attention from both industry and academia. Compared with emotion classification [1–4], there are many challenges for identifying emotion cause since it needs the deep semantic understanding of text and accurate location of emotion cause. In recent years, Gui et al. [5] take emotion cause recognition as a question answer problem, and then present a framework which includes a query (emotion word) and document (candidate clause) as the primary input. Then a co-attention neural network is proposed by Li et al. [6] to identify the emotion cause with emotion clause around the emotion word. The existing methods mainly focus on the emotion word and their clause.

Y. Chang and X. Zhu (Eds.): CCIR 2022, LNCS 13819, pp. 1–15, 2023.
https://doi.org/10.1007/978-3-031-24755-2_1

However, they neglect a significant fact that the position information is an emotion cause clue. Meanwhile, it also needs to mine the deep semantic relation between emotion clause and candidate clause by an effective way.

Exp1: 那天上午收到了offer, 我太激动了，期待上班的那一天。

Exp1: I got the offer in that forenoon (c1). I felt very excited (c2), looking forward to the day of work (c3).

For example, a post is published by a microblog user. This sample is composed of three clauses. It is obvious that the clause c2 is emotion clause which concludes the emotion word "excited". Especially, the clause c1 is emotion cause. Moreover, there are still many difficulties to recognize the certain emotion cause for the previous works [5, 6] by just introducing the emotion word, or producing the emotion clause. Moreover, the words "got" and "offer" can also be capable of motivating the emotion word "excited" with long distance. Therefore, the relation between emotion clause and candidate clause with position information can serve as a critical clue for recognizing emotion cause.

In this paper, we present a novel end-to-end position-aware word-level and clause-level attention network model (PWCA) to recognize the emotion cause. In addition to introduce the position information, PWCA also mutually builds the semantic relationship between emotion clause and candidate clause by adopting a word-level and clause-level attention mechanism. To be specific, our model is composed of two components: (1) According to obtaining position information of each clause based on the emotion clause, we convert the position information into position embedding. (2) PWCA builds the word-level and clause-level attention based on bidirectional GRU model to extract word-level and clause-level features. We evaluate our method on public emotion cause dataset, and the results demonstrate that our model is more effective than other previous approaches.

In brief, the main contributions of our work can be summarized as follows.

We attempt to explicitly investigate the effectiveness of the position information between emotion clause and candidate clause for emotion cause recognition.

We propose a position-aware word-level and clause-level attention network (PWCA) model based on Bi-GRU, which improves the performance of emotion cause recognition.

Experimental results demonstrate that our method can obviously improve the performance (measured by F1 score) of state-of-the-art model by 3.24% based on a released emotion cause dataset, and then further examples and attention visualization verifies the effectiveness of our proposed model for detecting emotion cause.

The rest of this paper is structured in the following four parts. Section 2 mainly reviews the related work on emotion cause recognition. Section 3 presents our proposed model for emotion cause recognition. Section 4 shows our detailed experiments and discusses evaluation results. Finally, Sect. 5 concludes our research contributions and offers the future work.

2 Related Work

Emotion cause extraction is capable of expressing the important semantic information, which text produces a complex emotion in an event and why element changes an emotion. In this section, we will briefly review some research on emotion cause in recent years, including emotion classification and emotion cause recognition approaches.

As we know, supervised machine learning methods and deep learning networks have been explored by adopting many different types of features and models for emotion classification. An unsupervised method was applied to control the emotion feeler and topic that proposed by Das and Bandyopadhpay [7]. Gao et al. [1] presented a joint learning model, which introduced an emotion classifier and polarity classifier. Chang et al. [8] designed many various types of linguistic features to settle with emotion classification. Cheng et al. [9] proposed an end-to-end hierarchical attention framework. In order to better understand the nuances, Barbieri et al. [10] leveraged a label-wise attention mechanism for emoji prediction. Moreover, Wang et al. [2] built a RNN-Capsule model to output word and reflect the attributes without adopting any other linguistic knowledge, which depended on Recurrent Neural Network. Gu et al. [11] employed a position-aware bidirectional attention network for emotion analysis. Wang et al. [12] adopt a sentence-level discourse segmentation to segment the sentences, and then modeled a hierarchical attention network for emotion classification.

In recent years, both industry and academia have realized the importance of recognizing and analyzing emotion cause. Lee et al. [13] first gave the definition of emotion cause recognition task, and then established a public balanced Chinese emotion cause dataset. Moreover, a knowledge base was built by Russo et al. [14] which related to emotion cause analysis. Gui et al. [15] designed many complied rules with labor cost to propose a rule-based emotion cause recognition model. Meanwhile, machine learning methods were usually applied to recognize the emotion cause in text by designing effective features, such as conditional random fields (CRF [16]) and Multi-Kernel Support Vector Machine (SVM, Gui et al. [17] and Xu et al. [18]). However, the current approaches are labor-intensive and highly depend on the quality of the features.

Compared with these above approaches, neural network architectures are able to learn important features combining with attention mechanism (Bahdanau et al.[19]), and have been widely leveraged in a variety of NLP tasks. Unexpectedly, there are many neural network models to recognize the emotion cause. Gui et al. [5] proposed novel deep question answer architecture by applying the deep convolutional memory network. Chen et al. [3] explored a joint deep learning framework in order to extract mutual benefits between two sub-tasks of emotion classification and emotion cause recognition. A co-attention deep learning network considers the emotion clause information based on Bi-LSTM and convolutional layer, which was proposed by Li et al. [6] for emotion cause analysis. Diao et al. [28] imported emotion word embeddi ng, synonyms embedding as enhanced representation, clause understanding layer and attention mechanism to recognize emotion. Bostan and Klinger [29] used the token sequence labeling to handle with English emotion stimulus detection task. Chen et al. [30] fully imported the conditional causal relationships between emotions and causes from text. Base on the graph, Hu et al. [31] proposed a graph convolutional network over the inter-clause dependency to fuse the semantics and structural information. Li et al. [32] formalized the emotion cause analysis as a sequence tagging task to extract specific emotion cause span in the given context. On the contrary, they ignore the position information and semantic relationship between emotion clause and candidate clause. To address this issue, we present the Position-aware Word-level and Clause-level Attention Network (PWCA) model to recognize the emotion cause by both adopting position information, word-level and clause-level features.

3 Methodology

In this section, we first describe the definition of emotion cause recognition task. Then, we introduce the presented model position-aware word-level and clause-level attention network (PWCA) for emotion cause recognition in detail. At last, we give the training details.

3.1 The Definition of Emotion Cause Recognition

The goal of this emotion cause recognition is to detect which clause introduces the emotion expression in text, which has given a document $D = \{c1, c2, \ldots cn\}$ about an emotion cause event concluding n clauses. For convenience, a certain emotion word and corresponding emotion cause have been included in each event. Existing works usually introduce emotion word as a query to recognize emotion cause [5], or just consider emotion clause as a query [6]. However, it is necessary to produce the position information and semantic relationship between emotion clause and candidate clause as an important clue for recognizing emotion cause as following.

3.2 Position-Aware Word-Level and Clause-Level Attention Network for Emotion Cause Recognition

In this section, we describe an end-to-end model position-aware word-level and clause-level attention network (PWCA) for emotion cause recognition, which is composed of four different layers as shown in Fig. 1: (1) an embedding layer for the input words combining with emotion word and position information; (2) position-aware word-level attention understanding layer for extracting informative words considering with word encoder layer; (3) position-aware clause-level attention understanding layer for producing attractive clauses taking with clause encoder layer; (4) classification layer for assigning the probability distribution of candidate clause. The architectures of our PWCA model are shown in Fig. 1. We will describe the details of the above four modules as follows.

Embedding Layer

Word Representation Module. Concretely, we firstly obtain the representation of each word and formalize the notations in our work. For each word wi, we derive its dense semantic embedding $v_i \in \mathbb{R}^d$, which can be initialized from pre-trained word embedding models (Mikolov et al. [20]) and subsequently adjusted during training.

Position Representation Module. As for how to introduce the position information of emotion clause with respect to its corresponding clause, inspired by the position encoding vectors proposed in (Zeng et al. [21]). Suppose that if this clause is the same as emotion

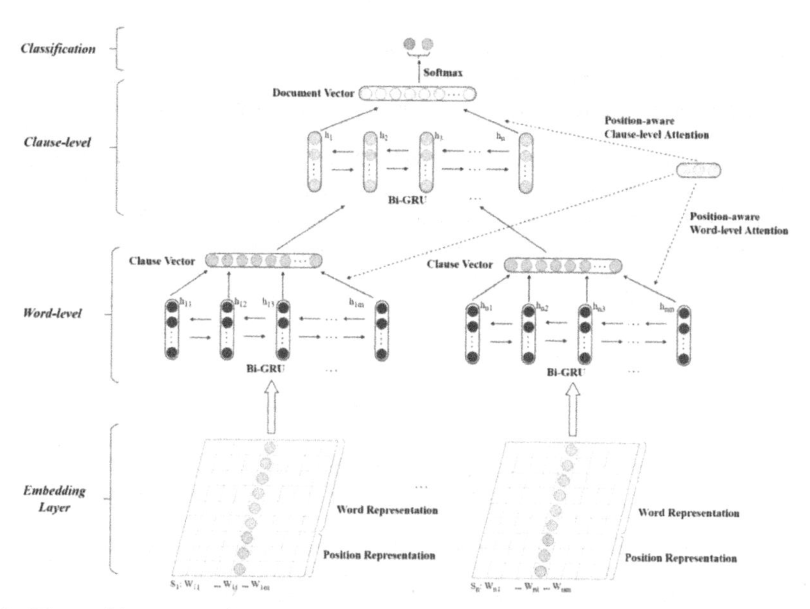

Fig. 1. The architectures of position-aware word-level and clause-level attention network (PWCA) model

clause, then its position index will be set to "0". Then the position index of other clauses will be marked as the relative distance to the current emotion clause.

$$P_i = \begin{cases} 0, & i = j \\ |i - j|, & i \neq j \end{cases} \tag{1}$$

where j denotes the position of emotion clause, and p_i can be viewed as the relative distance of candidate clause to emotion clause.

Thus, the corresponding position representation is generated by looking up a position embedding matrix $P \in \mathbb{R}^{d_p}$, which is randomly initialized and updated during the training process. Here, d_p represents the dimension of the embedding. By converting the embedding of position information, it is obvious that position embedding is able to build the different weights of words with different distance.

For each event, it still contains emotion word to express their opinion. We map the emotion word into high-dimensional vector space $e_i \in \mathbb{R}^{d'}$. Then, inspired by Wang et al. [22], we append the emotion representation and position representation to the embedding of each word to form an emotion-augmented and position-augmented embedding for each word.

$$X_i = v_i \oplus e_i \oplus p_i \tag{2}$$

where x_i denotes the concatenate operator $\oplus$ as the inputs of word-level attention understanding layer combining with emotion information e_i and position information p_i. Noted that the dimension of is $(d_i + d_e + d_p)$.

(2) Position-aware Word-level Attention Understanding Layer.
The goal of this layer is to extract the informative word-level features by adopting word encoding and position-aware word-level attention mechanism.

Word Encoding Module. In order to encode the entire sequence of word embedding, we leverage bidirectional gated recurrent units (Bi-GRU) with GRU cell Chung et al. [23], which contains a current input xt, previous hidden state ht−1 and the current hidden state ht = GRU (xt, ht−1) at each step t to model the temporal interactions between words in context. The formulas are as following.

$$Z_t = \sigma(W_z x_t + U_z h_{t-1} + b_z) \tag{3}$$

$$r_t = \sigma(W_r x_t + U_r h_{t-1} + b_r)$$

$$\tilde{h}_t = \tanh\left(W_h x_t + r_t^{\circ} U_r h_{t-1} + b_h\right)$$

$$h_t = z_t^{\circ} h_{t-1} + (1 - z_t)^{\circ} h_t$$

Then, we apply Bi-GRU, which can efficiently make use of past features and future features for a specific time framework, for obtaining annotations of words by producing information from both two directions of words. The Bi-GRU can be calculated as $h^w = \left[\overrightarrow{h_t}, \overleftarrow{h_t}\right]$, where one is from left direction to right $\overrightarrow{h_t} = GRU_{ltr}(x_t, \overrightarrow{h_{t-1}})$, another is from right to left $\overleftarrow{h_t} = GRU_{rtl}(x_t, \overleftarrow{h_{t-1}})$, then the two parts are concatenated to summarize the information of the whole clause by word-level features.

Similarly, the current candidate clause understanding vector h^e is also produced for candidate clause information at the same way, which input is the candidate clause representation combining with emotion embedding and position embedding.

Position-Aware Word-Level Attention Module. As we know, all words in a clause are not provided the equal information. Some words are more attractive and meaningful towards emotion cause recognition. Therefore, our model will distribute a weight of each word corresponding to its importance, when generating the representation of a clause from words. In order to learn and assign these weights among words in a clause, we explore a position-aware word-level attention mechanism to capture the informative word-level specific signal for emotion clause based on position information. The formulas are as follows.

$$U_t = (V_w)^T \tanh\left(W_w \cdot [h^w; h^e] + b_w\right)$$

$$\alpha_{ti} = \mathrm{softmax}(u_t) = \frac{\exp(u_{ti})}{\sum_j \exp(u_{tj})} \tag{4}$$

$$S_t^w = \sum\nolimits_i \alpha_{ti} h_i^w$$

where h^w is the word encoding vector, h^e is the candidate clause vector, W_w is the weight matrix and b_w is the bias term in the training process, V_w is weight vector and $(V_w)^T$ denotes its transpose, S_t^w is the hidden state clause vector among word encoding module for emotion clause to extract word-level features.

(3) Clause-level Attention Understanding Layer
This layer aggregates the word-level features into clause-level representation by adopting position-aware attention mechanism, which assigns the significant weights to the more salient clauses.

Clause Encoding Module. Here, we also leverage this candidate clause understanding vector h^e in order to incorporate the contextual information for capturing clause-level features. Meanwhile, context among the candidate clause is a significant clue to recognize the emotion cause. Therefore, we introduce the entire clause vectors c_i with current candidate clause c_t into the contextual clause-level vectors hc_i by Bi-GRU as the inputs of clause-level attention module.

Position-Aware Clause-Level Attention Module. Absolutely, different clauses can contribute unequally to semantics for emotion clause. Hence, in clause-level, we also employ a position-aware attention mechanism with candidate clause vector h^e to produce the clause representation. And hc_i is the annotation vector for clauses. By the contextual clause representations with position information, we obtain the attention weight between each clause vector hci and candidate clause vector h^e as follows.

$$M_i = (V_c)^T \tanh\left(W_c \cdot [hc_i; h^e] + b_c\right)$$

$$\alpha_i = \mathrm{softmax}(m_i) = \frac{\exp(m_i)}{\sum_j \exp(m_j)} \tag{5}$$

$$z = \sum_i \alpha_i hc_i$$

where V_c is weight vector and $(V_c)^T$ is the transposition, W_c is an intermediate matrix and b_c is an offset value which update in the learning process, z is the clause-level document representation based on the attention vectors α_i.

Classification Layer. Specifically, classification layer is used to calculate the whole possible labels, which depends on the word-level and clause-level features with position-aware information by previous layers. To perform emotion cause recognition, we feed the clause-level representation z to a softmax classifier with a linear transformation as follows.

$$O = \mathrm{softmax}(W_o \cdot z + b_o) \tag{6}$$

where W_o and b_o is the parameters of output layer, o denotes the predicted probability distribution for emotion cause. Finally, the label with the highest probability stands for the predicted distribution of emotion cause.

3.3 Model Training

In our model, PWCA can be trained in an end-to-end way in a supervised training architecture. The aim of the learning is to optimize all the parameters so as to minimize the objective function (loss function) as much as possible. Meanwhile, let y_i^j be the correct emotion cause which is represented by one-hot vector, and $\widehat{y_i}^j$ denotes the predicted probability distribution for the given clause. Therefore, the cross-entropy error between ground truth distributions and predicted distributions of emotion cause as the loss function, and the formula is as bellows.

$$\text{Loss} = -\sum_i \sum_j y_i^j \log \widehat{y_i}^j + \frac{1}{2}\lambda \|\theta\|^2 \quad (7)$$

where I is the index of clause, j is the index of class. λ is the regularization factor. θ contains all the training parameters.

Here, our proposed model can be trained by adopting stochastic gradient decent (SGD) methods. We also apply dropout strategy and early stopping strategy for settling with overfitting in order to enhance our PWCA model. More details will be described in the following experiments.

4 Experiment

In this section, we conduct experiments on real-world dataset to validate the effectiveness and advancement of our proposed PWCA model and report empirical results.

4.1 Experimental Settings

Dataset. To evaluate our presented model, the experiments are conducted on a public Chinese emotion cause corpus[1] [17] for understanding emotion cause. This dataset is composed of 2105 events which source is SINA Society new[2]. Particularly for each event, there is only each corresponding emotion word and one or more emotion causes. Here, we leverage CoreNLP[3] (Manning et al. [24]) tool for segmenting each event into a lot of clauses. The goal of emotion cause recognition is mainly to identify the emotion cause of each event. All statistics are shown in Table 1.

Table 1. Statistics of the dataset

Item	Number
Events	2105
Clauses	11799
Emotion Causes	2167
Events with 1 emotion	2046
Events with 2 emotions	56
Events with 3 emotions	3

[1] http://hlt.hitsz.edu.cn/?page_id=694.
[2] http://news.sina.com.cn/society/.
[3] http://stanfordnlp.github.io/CoreNLP/.

Word Embedding. In our experiments, all word embeddings are initialized by the pre-trained Word2Vec [20], which select Twitter and Sina Microblog as the unlabeled corpus. The dimension is set to 300.

Hyper-parameters. With this design, we use 10-fold cross validation to produce the best performance of the whole parameters on training dataset. Furthermore, out-of-vocabulary words are initialized by sampling from uniform distribution $U(-0.01, 0.01)$. All the weight matrices are given the initial value by sampling from the same distribution, and biases are set to zero. The dimension of word embedding is 300, and the dimension of position embedding is set to 20 which is randomly initialized and updated during the training procedure. Dropout strategy is also applied for avoiding overfitting and set to 0.5. The mini-batch is 32. The parameters of this model are optimized by adopting Adagrad [25].

Evaluation Metric. The standard metric is adopted in our experiments followed by Lee et al. [13], such as Precision, Recall and F1-measure. Furthermore, they are commonly accepted so that we can compare with other strong baselines.

4.2 Experimental Results

In this subsection, we give some baseline approaches for comparison in order to comprehensively evaluate the performance of our proposed PWCA model. Note that these existing methods are divided into three types: rule-based methods, machine learning methods, and neural network methods. Meanwhile, all the approaches employ the same representation for fair comparison.

Rule Based Methods. RB: This approach is a rule based approach [13], which designs many effective linguistic rules in order to identify emotion cause. CB: This approach is common sense knowledge based method [14], and relies on a Chinese Emotion Cognition Lexicon knowledge base.

Machine Learning Methods. RB + CB + ML: This approach only applies the designed features and linguistic rules combining with common sense knowledge base to detect emotion cause by a SVM classifier [26]. SVM: This approach captures the unigram, bigram and trigram features [27]. Word2Vec: This approach only employs word embedding features by adopting SVM classifier. Multi-kernel: This approach extends a multi-kernel function to recognize the emotion cause by SVM classifier [17].

Neural Network Methods. CNN: This approach only uses a basic deep convolutional neural network to detect the emotion cause. Bi-LSTM: This approach models the context from forward and backward directions based on a bidirectional long short time memory network. ConvMS-Memnet: This approach builds a deep memory network [5], which considers the convolutional information with three layers. CANN: This approach introduces a novel co-attention neural network framework in order to identify emotion cause. This is a state-of-the-art approach proposed by Li et al. [6]. EACN: This enhanced-representation attention convolutional-context network imports emotion word embedding, synonyms embedding, clause understanding layer and attention mechanism [28]. GCN: GCN model denotes the method proposed in [31]. AGGCN: AGGCN model denotes the method proposed in [31]. FSS-GCN: FSS-GCN model is a graph convolutional networks with fusion of semantic and structural information [31]. Absolutely, the best performances are chosen and shown from their paper.

Table 2 shows the performance comparison of the above different approaches, and we can find that.

Table 2. Comparison with previous approaches. The results with superscript * are reported in [6, 28] and [31]. The best results in each type are highlighted.

Methods	P (%)	R (%)	F1 (%)
RB*	67.47	42.87	52.43
CB*	26.72	71.30	38.87
RB + CB*	54.35	53.07	53.70
RB + CB + ML*	59.21	53.07	55.97
SVM*	42.00	43.75	42.85
Word2vec*	43.01	42.33	41.36
Multi-kernel*	65.88	69.27	67.52
CNN*	53.07	64.27	58.07
Bi-LSTM	61.45	61.95	61.70
ConvMS-Memnet*	70.76	68.38	69.55
CANN*	77.21	68.91	72.66
EACN*	72.15	73.76	72.94
GCN*	70.28	64.29	67.08
AGGCN*	76.27	69.30	72.50
FSS-GCN(layer1)*	73.66	73.51	73.46
PWCA	76.44	75.37	75.90

(1) Compared with RB (obtaining high precision but low recall) and CB (obtaining high recall but low precision), it is obvious that RB + CB has the better performance for identifying emotion cause, which denotes the effectiveness and advancement of combination with linguistic rule and common sense knowledge base.
(2) In the methods of machine learning, RB + CB + ML outperforms SVM and Word2vec by F1-measure. It verifies that the features and common sense can capture the effective semantic information combining with a machine learning classifier. Furthermore, multi-kernel method performs the best for all the machine learning methods by F1-measure. Because this extend method is capable of introducing the structured syntactic information to understand the emotion context.

(3) For neural network methods, we can find that the standard deep learning models are able to achieve the similar performance with multi-kernel, such as CNN and Bi-LSTM. The reason is that neural network model contains the ability of understanding latent semantic information. As, from all the results we can observe that the best F1 value is obtained by our presented PWCA model, which outperforms the CANN by 3.24%, outperforms the ConvMS-Memnet by 6.35%, outperforms the GCN by 8.82%, outperforms the AGGCN by 8.823.4% and outperforms the FSS-GCN by 2.56%. These results show that position information between candidate clause and emotion clause is a significant clue to recognize emotion cause. We can also find that our proposed PWCA model outperforms CNN by 17.83% and Bi-LSTM by 14.20% in F1-measure.

It verifies that the important positive effectiveness of our word-level and clause-level attention mechanism is introduced the hierarchical mutual relationship with emotion clause in order to improve the performance of emotion cause recognition.

4.3 Qualitative Analysis

We conduct extra experiments to do detailed analysis for this task as following.

Effect of Different PWCA Components
In order to investigate the effects of different components for our proposed components, we also implement independent hierarchical Bi-GRU network (HBi-GRU), hierarchical word-level attention network (HWA), hierarchical clause-level attention network (HCA), and position-aware word-level and clause-level attention network (PWCA) in this subsection. Table 3 demonstrates the performance of single attention mechanism with word-level or clause-level features. From the table, we can observe that:

Table 3. Effect of different PWCA components

Methods	P (%)	R (%)	F1 (%)
HBi-GRU	69.24	68.10	68.67
HWA	71.70	68.73	70.18
HCA	72.68	71.29	71.98
PWCA	76.44	75.37	75.90

(1) Compared to the model HBi-GRU only adopting local semantic information, HWA and HCA both achieve certain improvements, which verifies the rationality of incorporating word-level and clause-level features into emotion cause recognition via attention mechanism. It also shows that word-level attention or clause-level attention combining with position information can capture more semantic relationship of emotion cause.

(2) The clause-level information has the better performance than word-level information to enhance the clause representation. This is because contextual information among clauses can provide more significant clues to locate the emotion cause. Therefore, it is reasonable that the discrimination of clause-level features is more obvious than word-level characteristic.
(3) Our proposed PWCA model obtains the best performance compared to single word-level attention or clause-level attention, which shows that the information is able to contribute to our model. The results demonstrate our presented PWCA model is capable of extracting word-level and clause-level characteristic.

Effect of Fine-Grained Factors
In order to exploit the impacts of different fine-grained factors, we conduct the experiments for recognizing emotion cause in the same parameter setting. Results are demonstrated in Table 4.

Table 4. Effect of fine-grained factors

Methods	F1 (%)
No word embedding	63.18
No emotion word embedding	68.42
No position embedding	70.18
All	75.90

As shown in Table 4, we can find that: (1) Compared to random initialization, word embedding enables our model to provide the semantic space into represent the vector. (2) Emotion word embedding can provide the clues to capture the emotion cause since emotion word is able to affect the emotion context. (3) Position embedding is an important factor because emotion cause is mostly located to the emotion word. (4) Certainly, it is clear that the partial models cannot compete with the All model, which shows that our proposed PWCA model can improve the good performance by word-level and clause-level attention with position-aware information.

Effect of Different Attention Modules
To validate the effects of single word-level or clause-level attention, we replace average pooling operation to word-level attention or clause-level attention. Table 5 shows the performance of different attention modules.

From this Table 5, we can see that: (1) Average pooling has the better performance than only two-layer Bi-GRU (69.38% vs 68.67%), which shows the effectiveness of pooling strategy. (2) Both word-level attention and clause-level attention outperform average pooling. It verifies that attention mechanism is suitable to handle with natural language processing tasks. (3) When word-level and clause-level attention is leveraged, our approach outperforms all the state-of-the-art methods. These results encourage to

Table 5. Impact of fine-grained factors

Word-level	Clause-level	F1 (%)
Average pooling	Average pooling	69.38
Attention	Average pooling	71.29
Average pooling	Attention	72.53
Attention	Attention	75.90

incorporate the importance degrees of both word-level and clause-level features for recognizing the emotion cause.

5 Conclusion and Future Work

IN this paper, we present a hierarchical position-aware attention model for emotion cause recognition. The main idea of our proposed PWCA model is to leverage both word-level and clause-level attention combining with position-aware information and emotion information to incorporate the knowledge of word-level and clause-level features. Experimental results on a real-world corpus demonstrate that our approach outperforms a lot of competitive baselines and show the effectiveness and advancement of our proposed PWCA model.

In our future work, we would like to apply more information for identifying emotion cause, such as semantic relationships between clauses to improve the classification performance. Moreover, we will also leverage our word-level and clause-level architecture to settle with other sentiment analysis tasks.

Acknowledgments. This work is partially supported by grant from the Natural Science Foundation of China (No. 62006130, 62066044), Inner Mongolia Science Foundation (No. 2022MS06028).

References

1. Gao, W., Li, S., Yat, S., Lee, M., Zhou, G., Huang, C.R.: Joint learning on sentiment and emotion classification. In: CIKM, pp. 1505–1508 (2013)
2. Wang, Y., Sun, A., Han, J., et al.: Sentiment analysis by capsules. In: Proceedings of the 2018 World Wide Web Conference on World Wide Web. International World Wide Web Conferences Steering Committee, pp. 1165–1174 (2018)
3. Chen, Y., Hou, W., Cheng, X., et al.: Joint learning for emotion classification and emotion cause detection. In: Proceedings of the 2018 Conference on Empirical Methods in Natural Language Processing, pp. 646–651 (2018)
4. Truong, Q.T., Lauw, H.W.: VistaNet: visual aspect attention network for multimodal sentiment analysis. In: AAAI (2019)
5. Gui, L., Hu, J., He, Y., et al.: A question answering approach to emotion cause extraction. In: EMNLP (2017)

6. Li, X., Song, K., Feng, S., et al.: A co-attention neural network model for emotion cause analysis with emotional context awareness. In: Proceedings of the 2018 Conference on Empirical Methods in Natural Language Processing, pp. 4752–4757 (2018)
7. Das, D., Bandyopadhyay, S.: Emotions on Bengali blog texts: role of holder and topic. In: International Conference on Advances in Social Networks Analysis and Mining, pp. 587–592 (2011)
8. Chang, Y.C., Chen, C.C., Hsieh, Y.L., Hsu, W.L.: Linguistic template extraction for recognizing reader-emotion and emotional resonance writing assistance. In: ACL, pp. 775–780 (2015)
9. Cheng, J., Zhao, S., Zhang, J., et al.: Aspect-level sentiment classification with heat (hierarchical attention) network. In: Proceedings of the 2017 ACM on Conference on Information and Knowledge Management, pp. 97–106. ACM (2017)
10. Barbieri, F., Anke, L.E., Camacho-Collados, J., et al.: Interpretable emoji prediction via label-wise attention LSTMs. In: Proceedings of the 2018 Conference on Empirical Methods in Natural Language Processing, pp. 4766–4771 (2018)
11. Gu, S., Zhang, L., Hou, Y., et al.: A position-aware bidirectional attention network for aspect-level sentiment analysis. In: Proceedings of the 27th International Conference on Computational Linguistics, pp. 774–784 (2018)
12. Wang, J., Li, J., Li, S., et al.: Aspect sentiment classification with both word-level and clause-level attention networks. In: IJCAI, pp. 4439–4445 (2018)
13. Lee, S.Y.M., Chen, Y., Huang, C.R.: A text-driven rule-based system for emotion cause detection. In: Proceedings of the NAACL HLT 2010 Workshop on Computational Approaches to Analysis and Generation of Emotion in Text. Association for Computational Linguistics, pp. 45–53 (2010)
14. Russo, I., Caselli, T., Rubino, F., Boldrini, E., Mart´ınez-Barco, P.: Emocause: an easy-adaptable approach to emotion cause contexts. In: Workshop on Computational Approaches to Subjectivity and Sentiment Analysis, pp. 153–160 (2011)
15. Gui, L., Yuan, L., Xu, R., et al.: Emotion cause detection with linguistic construction in chinese weibo text. In: Natural Language Processing and Chinese Computing, pp. 457–464. Springer, Heidelberg (2014)
16. Ghazi D, Inkpen D, Szpakowicz S. Detecting emotion stimuli in emotion-bearing sentences. International Conference on Intelligent Text Processing and Computational Linguistics. Springer, Cham, 2015: 152–165
17. Gui, L., Wu, D., Xu, R., Lu, Q., Zhou, Y.: Event-driven emotion cause extraction with corpus construction. In: EMNLP, pp. 1639–1649 (2016)
18. Xu, R., Hu, J., Lu, Q., et al.: An ensemble approach for emotion cause detection with event extraction and multi-kernel SVMs. Tsinghua Sci. Technol. **22**(6), 646–659 (2017)
19. Bahdanau, D., Cho, K., Bengio, Y.: Neural machine translation by jointly learning to align and translate. Comput. Sci. (2014)
20. Mikolov, T., Chen, K., Corrado, G., et al.: Efficient estimation of word representations in vector space. In: ICLR, pp. 1–12 (2013)
21. Zeng, D., Liu, K., Lai, S., Zhou, G., Zhao, J.: Relation classification via convolutional deep neural network (2014)
22. Wang, Y., Huang, M., Zhao, L.: Attention-based lstm for aspect-level sentiment classification. In: Proceedings of the 2016 Conference on Empirical Methods in Natural Language Processing (2016)
23. Chung, J., Gulcehre, C., Cho, K.H., et al.: Empirical evaluation of gated recurrent neural networks on sequence modeling (2014). arXiv preprint arXiv:1412.3555
24. Manning, C., Surdeanu, M., Bauer, J., et al.: The Stanford CoreNLP natural language processing toolkit. In: Proceedings of 52nd Annual Meeting of the Association for Computational Linguistics: System Demonstrations, pp. 55–60 (2014)

25. Duchi, J., Hazan, E., Singer, Y.: Adaptive subgradient methods for online learning and stochastic optimization. J. Mach. Learn. Res. **12**, 2121–2159 (2011)
26. Chen, Y., Lee, S.Y.M., Li, S., Huang, C.R.: Emotion cause detection with linguistic constructions. In: Coling, pp. 179–187 (2010)
27. Li, W., Xu, H.: Text-based emotion classification using emotion cause extraction. Expert Syst. Appl. **41**(4), 1742–1749 (2014)
28. Diao, Y., Lin, H., Yang, L., et al.: Emotion cause detection with enhanced-representation attention convolutional-context network. Soft. Comput. **25**(2), 1297–1307 (2021)
29. Bostan, L.A.M., Klinger, R.: Token sequence labeling vs. clause classification for english emotion stimulus detection. In: Proceedings of the Ninth Joint Conference on Lexical and Computational Semantics, COLING 2020, Barcelona, Spain, 12–13 December 2020, pp. 58–70. Association for Computational Linguistics (2020)
30. Chen, X., Li, Q., Wang, J.: Conditional causal relationships between emotions and causes in texts. In: Proceedings of the 2020 Conference on Empirical Methods in Natural Language Processing, EMNLP 2020, Online, 16–20 November 2020, pp. 3111–3121. Association for Computational Linguistics (2020)
31. Hu, G., Lu, G., Zhao, Y.: FSS-GCN: a graph convolutional networks with fusion of semantic and structure for emotion cause analysis. Knowl.-Based Syst. **212**(1), 106584 (2020)
32. Li, X., Gao, W., Feng, S., et al.: Boundary detection with BERT for span-level emotion cause analysis. In: Findings of the Association for Computational Linguistics: ACL-IJCNLP 2021 (2021)

ID-Agnostic User Behavior Pre-training for Sequential Recommendation

Shanlei Mu[1], Yupeng Hou[2], Wayne Xin Zhao[2,4](✉), Yaliang Li[3], and Bolin Ding[3]

[1] School of Information, Renmin University of China, Beijing, China
slmu@ruc.edu.cn
[2] Gaoling School of Artifical Intelligence, Renmin University of China, Beijing, China
houyupeng@ruc.edu.cn, batmanfly@gmail.com
[3] Alibaba Group, Sunnyvale, USA
{yaliang.li,bolin.ding}@alibaba-inc.com
[4] Beijing Key Laboratory of Big Data Management and Analysis Methods, Beijing, China

Abstract. Recently, sequential recommendation has emerged as a widely studied topic. Existing researches mainly design effective neural architectures to model user behavior sequences based on item IDs. However, this kind of approach highly relies on user-item interaction data and neglects the attribute- or characteristic-level correlations among similar items preferred by a user. In light of these issues, we propose **IDA-SR**, which stands for **ID-A**gnostic User Behavior Pre-training approach for **S**equential **R**ecommendation. Instead of explicitly learning representations for item IDs, IDA-SR directly learns item representations from rich text information. To bridge the gap between text semantics and sequential user behaviors, we utilize the pre-trained language model as text encoder, and conduct a pre-training architecture on the sequential user behaviors. In this way, item text can be directly utilized for sequential recommendation without relying on item IDs. Extensive experiments show that the proposed approach can achieve comparable results when only using ID-agnostic item representations, and performs better than baselines by a large margin when fine-tuned with ID information.

Keywords: Recommender system · Sequential recommendation · User behavior modeling

1 Introduction

Over the past decade, sequential recommendation has been a widely studied task [9], which learns time-varying user preference and provides timely information resources in need. To better model sequential user behaviors, recent approaches [4,6,10,11] mainly focus on how to design effective neural architecture for recommender system, such as Recurrent Neural Network (RNN) [4] and Transformer [6,10]. Typically, existing methods learn the item representations

Y. Chang and X. Zhu (Eds.): CCIR 2022, LNCS 13819, pp. 16–27, 2023.
https://doi.org/10.1007/978-3-031-24755-2_2

based on item IDs (*i.e.*, a unique integer associated with an item) and then feed them to the designed sequential recommender. We refer to such a paradigm as *ID-based item representation.*

In the literature of recommender systems, ID-based item representation has been a mainstream paradigm to model items and develop recommendation models. It is conceptually simple and flexibly extensible. However, it is also noticed with several limitations for sequential recommendation. First, it highly relies on user-item interaction data to learn ID-based item representations [4,9]. When interaction data is insufficient, it is difficult to derive high-quality item representations, which are widely observed in practice [7,19]. Second, there is a discrepancy between real user behaviors and the learned sequential models. Intuitively, user behaviors are driven by underlying user preference, *i.e.*, the preference over different item attributes or characteristics, so that a user behavior sequence essentially reflects the attribute- or characteristic-level correlations among similar items preferred by a user. However, ID-based item representations learn more abstractive ID-level correlations, which cannot directly characterize fine-grained correlations that actually reflect real user preference.

Considering the above issues, we aim to represent items in a more natural way from the user view, such that it can directly capture attribute- or characteristic-level correlations without explicitly involving item IDs. The basic idea is to incorporate item side information in modeling item sequences, where we learn *ID-agnostic item representations* that are derived from attribute- or characteristic-level correlations driven by user preference. Specifically, in this work, we utilize rich text information to represent items instead of learning ID-based item representations. As a kind of natural language, item text reflects the human's cognition about item attributes or characteristics, which provides a general information resource to reveal user preference from sequential behavior. Different from existing studies that leverage item text to enhance ID-based representations [16,17] or improve zero-shot recommendations [3], this work aims to learn sequential user preference solely based on text semantics without explicitly modeling item IDs.

However, it is not easy to transfer the semantics reflected in the item texts to the recommendation task, as the semantics may not be directly relevant to the user preference or even noisy to the recommendation task [1]. There is a natural semantic gap between natural language and user behaviors. To fill this gap, we construct a pre-trained user behavior model that learns user preference by modeling the text-level correlations through behavior sequences.

To this end, we propose an **ID**-**A**gnostic User Behavior Pre-training approach for **S**equential **R**ecommendation, named **IDA-SR**. Compared with existing sequential recommenders, the most prominent feature of IDA-SR is that it no longer explicitly learn representations for item IDs. Instead, it directly learns item representations from item texts, and we utilize the pre-trained language model [2] as text encoder to represent items. To adapt text semantics to the recommendation task, the key point is a pre-training architecture conducted on the sequential user behaviors with three important pre-training tasks, namely *next item prediction*, *masked item prediction* and *permuted item prediction*. In this way, our approach can effectively bridge the gap between text semantics and sequential user behaviors, so that item text can be directly utilized for sequential recommendation without the help of item IDs.

To evaluate the proposed approach, we conduct extensive experiments on the four real-world datasets. Experimental results demonstrate that only using ID-agnostic item representations, the proposed approach can achieve comparable results with several competitive recommendation methods and even much better when training data is limited. In order to better fit the recommendation task, we also provide a fine-tuning mechanism that allows the pre-trained architecture to use the guidance of explicit item ID information. In such way, our approach performs better than baseline methods by a large margin under various settings, which is brought by the benefit of the ID-agnostic user behavior pre-training.

2 Preliminaries

In this section, we first introduce the used notations throughout the paper, and then formulate the sequential recommendation problem.

Notations. Assume that we have a set of users and items, denoted by $\mathcal{U}$ (size $|\mathcal{U}|$) and $\mathcal{I}$ (size $|\mathcal{I}|$), respectively, where $u \in \mathcal{U}$ denotes a user and $i \in \mathcal{I}$ denotes an item. Generally, a user u has generated a chronologically-ordered interaction sequence with items: $\{i_1, \cdots, i_n\}$, where n is the number of interactions and i_t is the t-th item that the user u has interacted with. For convenience, we use $i_{j:k}$ to denote the subsequence, *i.e.*, $i_{j:k} = \{i_j, \cdots, i_k\}$ where $1 \leq j \leq k \leq n$. For each item i, it is also associated with a *item text* describing its attributes or characteristics, denoted by $C_i = \{w_1, \cdots, w_m\}$, where w denotes a word from the vocabulary. In this work, we concatenate the item title, brand and category label as item texts. As will be seen later, our approach is general to incorporate various kinds of associated texts for modeling items.

Problem Statement. Based on the above notations, we define the task of sequential recommendation. Given the historical behaviors of a user $\{i_1, \cdots, i_n\}$ and the item text C_i for each item i, the task of sequential recommendation is to predict the next item that the user is likely to interact the $(n+1)$-th step.

3 Methodology

Given the ordered sequence of user u's historical items up to the timestamp-t: $\{i_1, \cdots, i_t\}$, we need to predict the next item, *i.e.*, i_{t+1}. In this section, we present the **ID-A**gnostic user behavior modeling approach for **S**equential **R**ecommendation, named **IDA-SR**. It consists of *ID-agnostic user behavior pre-training stage* and *fine-tuning stage*. The major novelty lies in the ID-agnostic user behavior pre-training, where we represent items solely based on item texts instead of item IDs. To adapt text semantics to the recommendation task, we incorporate an adapter layer that transforms text representations into item representations, and further design three elaborate pre-training tasks to retain the preference characteristics based on user behavior sequences. Figure 1 presents the overview of the pre-training stage. Next, we describe our approach in detail.

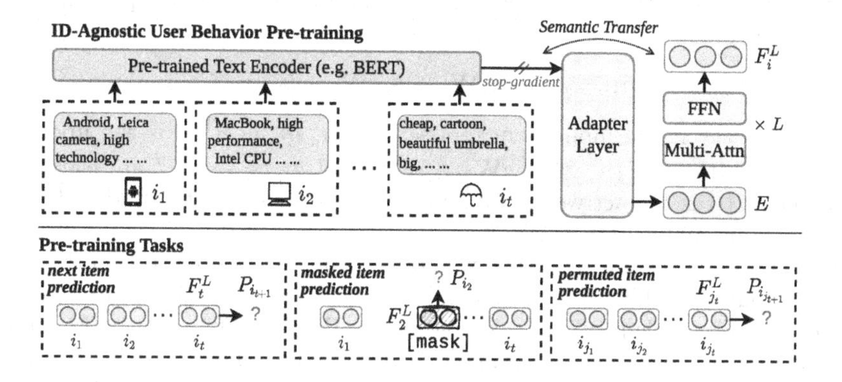

Fig. 1. Overall illustration of the proposed approach.

3.1 ID-Agnostic User Behavior Pre-training

For sequential recommendation, the key point of learning ID-agnostic representations is to capture sequential preference characteristics from user behavior sequences based on item texts. Since item texts directly describe items' attributes or characteristics, our pre-training approach tries to integrate item characteristic encoding into sequential user behavior modeling, and further bridge the semantic gap between them.

ID-Agnostic Item Representations. To obtain ID-agnostic item representations, our idea is to utilize the pre-trained language model (PLM) [2] to encode item texts. Specifically, we use the pre-trained language model BERT as the text encoder to generate item representations based on the item text. Given an item i and its corresponding item text $C_i = \{w_1, \cdots, w_m\}$, we first add an extra token [CLS] into the item text C_i to form the input word sequence $\tilde{C}_i = \{[\texttt{CLS}], w_1, \cdots, w_m\}$. Then the input word sequence $\tilde{C}_i$ is fed to the pre-trained BERT model. We use the embedding for the [CLS] token as ID-agnostic item representations. In this way, we can obtain the item text embedding matrix $\mathbf{M}_T \in \mathbb{R}^{|\mathcal{I}|\times\tilde{d}}$, where $\tilde{d}$ is the dimension of item text embedding.

Different from ID-based item representations, ID-agnostic item representations are less sensitive to the quality of the interaction data. Instead, it allows the sequential model to capture attribute- characteristics preference from user behavior sequences. It is also more resistible to the cold-start scenarios where a new item occurs for recommendation.

Text Semantic Adapter Layer. Although the ID-agnostic item representations generated from PLMs have great expressive ability for item characteristics, not all the encoded semantics in these representations are directly beneficial or useful for sequential user behavior modeling. Therefore, we incorporate a text semantic adapter layer for transforming the original text representations into a form that is more suitable for the recommendation task. The adapter layer is formalized as follows:

$$\tilde{\boldsymbol{m}}_i = \sigma\big(\sigma(\boldsymbol{m}_i\mathbf{W}_1 + \boldsymbol{b}_1)\mathbf{W}_2 + \boldsymbol{b}_2\big), \quad (1)$$

where $\boldsymbol{m}_i \in \mathbf{M}_T$ is the input item representation, $\tilde{\boldsymbol{m}}_i \in \mathbb{R}^{1\times d}$ is the updated item representation, $\mathbf{W}_1 \in \mathbb{R}^{\tilde{d}\times d}$, $\mathbf{W}_2 \in \mathbb{R}^{d\times d}$ and $\boldsymbol{b}_1, \boldsymbol{b}_2 \in \mathbb{R}^{1\times d}$ are learnable parameters, $\sigma(\cdot)$ is the activation function. So we can obtain the updated item text embedding matrix $\tilde{\mathbf{M}}_T \in \mathbb{R}^{|\mathcal{I}|\times d}$.

Given a n-length item sequence, we apply a look-up operation from $\tilde{\mathbf{M}}_T$ to form the input embedding matrix $\mathbf{E}_T \in \mathbb{R}^{n\times d}$. Besides, following [6], we further incorporate a learnable position encoding matrix $\mathbf{P} \in \mathbb{R}^{n\times d}$ to enhance the input representation of the item sequence. By this means, the sequence representations $\mathbf{E} \in \mathbb{R}^{n\times d}$ can be obtained by summing two embedding matrices $\mathbf{E} = \mathbf{E}_T + \mathbf{P}$.

Sequential User Behavior Modeling. The core of sequential recommendation lies in the sequential user behavior modeling, where we aim to capture sequential preference characteristics from user behavior sequences. Here, we adopt a classic self-attention architecture [12] based on the above ID-agnostic item representations. A self-attention block generally consists of two sub-layers, *i.e.*, a multi-head self-attention layer (denoted by MultiHeadAttn($\cdot$)) and a point-wise feed-forward network (denoted by FFN($\cdot$)). The update process can be formalized as following:

$$\mathbf{F}^{l+1} = \mathrm{FFN}(\mathrm{MultiHeadAttn}(\mathbf{F}^{l})), \quad (2)$$

where the $\mathbf{F}^l$ is the l-th layer's input. When $l = 0$, we set $\mathbf{F}^0 = \mathbf{E}$.

Text Semantics Based User Behavior Pre-training Task. Given the ID-agnostic item representations and self-attention architecture, we next focus on designing suitable optimization objectives to learn the parameters of the architecture, which is the key to bridge the semantic gap between text semantics and sequential preference characteristics. Next, we introduce three pre-training tasks derived from user behavior sequences based on text representations.

Next Item Prediction. The *next item prediction* pre-training task refers to predicting the next item having read all the previous ones, which has been widely adopted in the existing sequential recommendation methods [6]. Based on this task, we calculate the user's preference over the candidate item set as follows:

$$P_{pre}(i_{t+1}|S) = \mathrm{softmax}(\mathbf{F}_t^L\tilde{\mathbf{M}}_T^\top)_{[i_{t+1}]}, \quad (3)$$

where $S = i_{1:t}$ is the user historical sequence, $\mathbf{F}_t^L$ is the output of the L-layer self-attention block at step t. Such a task tries to capture sequential preference based on text semantics.

Masked Item Prediction. The *masked item prediction* pre-training task refers that corrupting the input item sequence and trying to reconstruct the original item sequence. Specifically, we randomly mask some items (*i.e.,* replace them with a special token `[MASK]`) in the input sequences, and then predict the masked items based on their surrounding context. Based on this task, we calculate the user's preference over the candidate item set as follows:

$$P_{pre}(i_t|S) = \text{softmax}(\mathbf{F}_t^L \tilde{\mathbf{M}}_T^\top)_{[i_t]}, \tag{4}$$

where S is the masked version for user behavior sequence, position t is replaced with `[MASK]`. Such a task enhances the overall sequential modeling capacity of the recommendation model.

Permuted Item Prediction. The *permuted item prediction* pre-training task refers that permuting the items in the original user behavior sequence, then using the previous items in permuted user behavior sequence to predict the next item. Given the input user behavior sequence $i_{1:t} = \{i_1, \cdots, i_t\}$, we first generate the permuted user behavior sequence $i_{j_1:j_t} = \{i_{j_1}, \cdots, i_{j_t}\}$ by permuting the items. Then, based on the permuted sequence, we calculate the user's preference over the candidate item set as follows:

$$P_{pre}(i_{j_{t+1}}|S) = \text{softmax}(\mathbf{F}_{j_t}^L \tilde{\mathbf{M}}_T^\top)_{[i_{j_{t+1}}]}, \tag{5}$$

where $S = i_{j_1:j_t}$ is the permuted user historical seuqence. In this way, the context for each position consist of tokens from both left and right, which is able to improve the performance [15].

To combine the three pre-training tasks, we adopt the cross-entropy loss to pre-train our model as follows:

$$\mathcal{L}_{pre} = -\sum_{u\in\mathcal{U}}\sum_{t\in\mathcal{T}} \log P_{pre}(i_t = i_t^*|S), \tag{6}$$

where i_t^* is the ground truth item, $\mathcal{T}$ is the predicted position set.

3.2 Fine-Tuning for Recommendation

At the pre-training stage, we integrate the text semantics of items into the sequential behavior modeling. Next, we further optimize the architecture according to the recommendation task. Different from previous self-supervised recommendation models [10,19], we can fine-tune our approach *with* or *without* item IDs.

Fine-Tuning Without ID. Without adding any extra parameters, we can directly fine-tune the pre-trained model based on the ID-agnostic item representations. In this way, we calculate the user's preference score for the item i in the step t under the context from user history as:

$$P_{fine}(i_{t+1}|i_{1:t}) = \text{softmax}(\mathbf{F}_t^L \tilde{\mathbf{M}}_T^\top)_{[i_{t+1}]}, \tag{7}$$

where $\tilde{\mathbf{M}}_T$ is the updated item text embedding matrix, $\mathbf{F}_t^L$ is the output of the L-layer self-attention block at step t. Since no additional parameters are incorporated, it enforces the model to well fit the recommendation task in an efficient way.

Fine-tuning with ID. Unlike text information, IDs are more discriminative to represent an item, *e.g.*, an item will be easily identified when we know its ID. Therefore, we can further improve the discriminative ability of the above pre-trained approach by incorporating additional item ID representations. Specifically, we maintain a learnable item embedding matrix $\mathbf{M}_I \in \mathbb{R}^{|\mathcal{I}|\times d}$. Then we combine $\mathbf{M}_I$ and the item text embedding $\tilde{\mathbf{M}}_T$ as the final item representation. We calculate the user's preference score for the item i in the step t under the context from user history as:

$$P_{fine}(i_{t+1}|i_{1:t}) = \text{softmax}(\tilde{\mathbf{F}}_t^L(\tilde{\mathbf{M}}_T + \mathbf{M}_I)^\top)_{[i_{t+1}]}. \tag{8}$$

Here, we only incorporate item IDs at the fine-tuning stage and the rest parts (ID-agnostic item representations, adapter layer, and self-attention architecture) have been pre-trained at the pre-training stage. As will be shown in Sect. 4.2, this fine-tuning method is more effective than that simply combining text and ID features.

For each setting, we adopt the widely used cross-entropy loss to train the model in the fine-tuning stage.

4 Experiments

4.1 Experimental Setup

Datasets. We conduct experiments on the Amazon review datasets [8], which contain product ratings and reviews in 29 categories on Amazon.com and rich textual metadata such as *title*, *brand*, *description*, etc. We use the version released in the year 2018. Specifically, we use the 5-core data of *Pantry*, *Instruments*, *Arts* and *Food*, in which each user or item has at least 5 associated ratings. The statistics of our datasets are summarized in Table 1.

Table 1. Statistics of the datasets.

Dataset	Pantry	Instruments	Arts	Food
# Users	13,101	24,962	45,486	115,349
# Items	4,898	9,964	21,019	39,670
# Actions	126,962	208,926	395,150	1,027,413

Comparison Methods. We consider the following baselines for comparisons:

- **PopRec** recommends items according to the item popularity.
- **FPMC** [9] models the behavior correlations by Markov chain.

- **GRU4Rec** [4] applies GRU to model user behaviors.
- **SASRec** [6] applies self-attention mechanism to model user behaviors.
- **BERT4Rec** [10] applies bidirectional self-attention mechanism to model user behaviors.
- **FDSA** [17] constructs a feature sequence and uses a feature level self-attention block to model the feature transition patterns.
- **ZESRec** [3] regards BERT representations as item representations for cross-domain recommendation. We report its result on source domain data.
- **S^3-Rec** [19] pre-trains user behavior models via mutual information maximization objectives for feature fusion.

Among all the above methods, FPMC, GRU4Rec, SASRec and BERT4Rec are general sequential recommendation methods that model the user behavior sequences only by user-item interaction data. FDSA, ZESRec and S^3-Rec are text-enhanced sequential recommendation methods that model the user behavior sequences with extra information from item texts. For our proposed approach, **IDA-SR$_t$** and **IDA-SR$_{t+ID}$** represent the model is fine-tuned without ID and with ID, respectively. We implement these methods by using the popular open-source recommendation library RecBole [18].

Evaluation Settings. To evaluate the performance, we adopt top-k Hit Ratio (HR@k) and top-k Normalized Discounted Cumulative Gain (NDCG@k) evaluation metrics. Following previous works [6,10], we apply the *leave-one-out* strategy for evaluation. For each user, we treat all the items that this user has not interacted with as negative items.

4.2 Experimental Results

In this section, we first compare the proposed IDA-SR approach with the aforementioned baselines on the four datasets, then conduct the ablation study, and finally compare the results on cold-start items.

Main Results. Compared with the general sequential recommendation methods, text-enhanced sequential recommendation methods perform better on some datasets. Because item texts are used as auxiliary features to help improve the recommendation performance. These results further confirm that semantic information from item text is useful for modeling user behaviors. By comparing the proposed approach IDA-SR$_{t+ID}$ with all the baselines, it is clear that IDA-SR$_{t+ID}$ consistently performs better than them by a large margin. Different from these baselines, we adopt the ID-agnostic user behavior pre-training framework, which transfers the semantic knowledge to guide the user behavior modeling through the three appropriate self-supervised tasks. Then in the fine-tuning stage, we fine-tune the model to utilize the user behavior knowledge from the pre-trained model. In this way, our proposed approach can better capture user behavior patterns and achieve much better results. Besides, without incorporating item ID information in the fine-tuning stage, IDA-SR$_t$ also has a comparable result with other baseline methods which model user behaviors based on the ID information. This further illustrates the effectiveness of our proposed approach.

Table 2. Performance comparison of different methods on the four datasets. The best performance and the best performance baseline are denoted in bold and underlined fonts respectively.

Method	Pantry		Instruments		Arts		Food	
	H@10	N@10	H@10	N@10	H@10	N@10	H@10	N@10
PopRec	0.0068	0.0024	0.0133	0.0039	0.0156	0.0090	0.0281	0.0141
FPMC	0.0373	0.0196	0.1043	0.0771	0.0958	0.0684	0.0940	0.0746
GRU4Rec	0.0395	0.0194	0.1045	0.0796	0.0909	0.0637	0.1075	0.0862
SASRec	0.0488	0.0231	0.1103	0.0787	0.1164	0.0685	0.1173	0.0846
BERT4Rec	0.0311	0.0160	0.1057	0.0697	0.1096	0.0774	0.1119	0.0792
FDSA	0.0422	0.0226	0.1117	0.0840	0.1074	0.0768	0.1124	0.0883
ZESRec	0.0529	0.0263	0.1076	0.0711	0.0971	0.0579	0.0967	0.0646
S^3-Rec	0.0509	0.0242	0.1123	0.0795	0.1093	0.0692	0.1163	0.0864
IDA-SR$_t$	0.0738	**0.0378**	0.1250	0.0821	0.1130	0.0708	0.1097	0.0730
IDA-SR$_{t+ID}$	**0.0750**	0.0375	**0.1304**	**0.0872**	**0.1304**	**0.0828**	**0.1309**	**0.0943**
Improv.	+41.78%	+43.73%	+16.12%	+3.81%	+12.03%	+6.98%	+11.59%	+6.80%

Ablation Study. We examine the performance of IDA-SR's variants by removing each pre-training task from the full approach. We use *np*, *mp* and *pp* to represent three pre-training task *next item prediction*, *masked item prediction* and *permuted item prediction*, respectively. Figure 2 presents the evaluation results. We can observe that the three pre-training tasks all contribute to the final performance. All of them are helpful for bridging the semantic gap between text semantics and sequential preference characteristics.

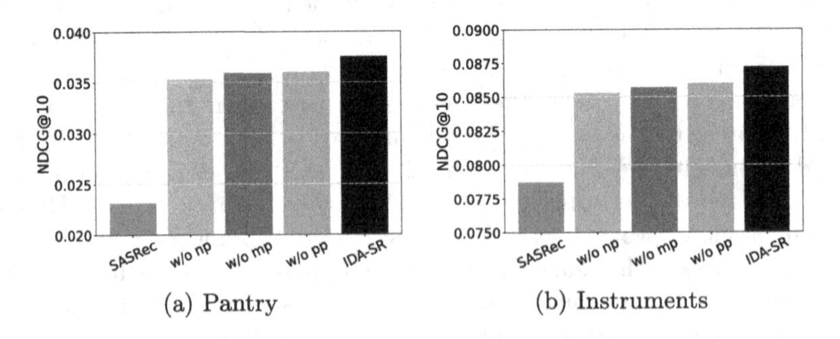

(a) Pantry (b) Instruments

Fig. 2. Ablation study.

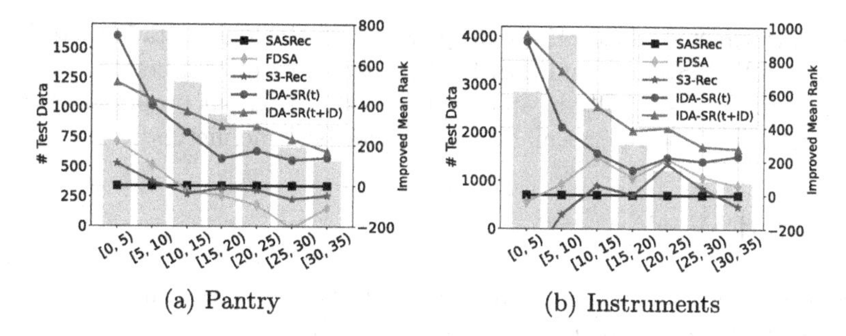

Fig. 3. Performance comparison w.r.t. cold-start items. The bar graph represents the number of test data for each group and the line chart represents the improved mean rank of ground truth item compared with SASRec.

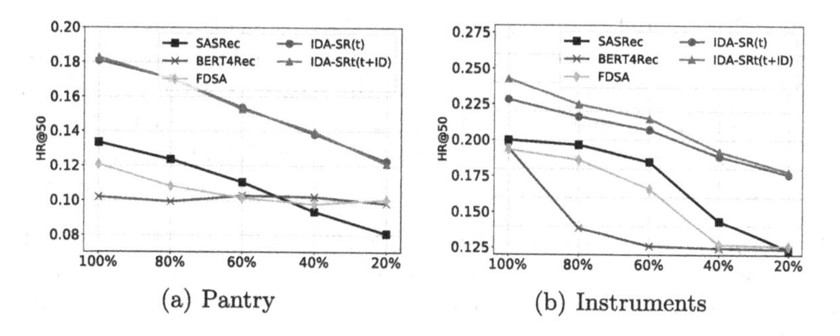

Fig. 4. Performance comparison w.r.t. different sparsity levels.

Performance Comparison w.r.t. Cold-Start Items. Conventional user behavior modeling methods are likely to suffer from the cold-start items recommendation problem. This problem can be alleviated by our method because the proposed ID-agnostic pre-training framework can utilize the text semantic information to make the model less dependent on the interaction data. To verify this, we split the test data according to the popularity of ground truth items in the training data, and then record the improved mean rank of ground truth item in each group compared with baseline method SASRec. From Fig. 3, we can find that the proposed IDA-SR$_t$ and IDA-SR$_{t+ID}$ have a big improvement when the ground truth item is extremely unpopular *e.g.*, group [0, 5) and group [5, 10). This observation implies the proposed IDA-SR can alleviate the cold-start items recommendation problem.

Performance Comparison w.r.t. Different Sparsity Levels. To further verify the proposed approach is less dependent on the interaction data and achieve better results than conventional recommendation methods, we examine how the proposed IDA-SR approach performs w.r.t. different sparsity levels. We simulate the data sparsity scenarios by using different proportions of the full datasets, *i.e.*, 100%, 80%, 60%, 40% and 20%. We summarize the experimental results on the Pantry and Instruments datasets in Fig. 4. From it, we can observe that the performance substantially drops when less training data is used. While our approach IDA-SR$_{t+ID}$ and IDA-SR$_t$ consistently perform better than other baseline methods. Especially in an extreme level (20%), our proposed approach can still achieve a good performance than other baseline methods.

5 Related Work

In this section, we review the related works in two aspects, namely sequential recommendation and pre-training for recommendation.

Sequential Recommendation. Sequential recommendation aims to predict the successive item(s) that a user is likely to adopt given her historical behaviors [4,9]. With the development of neural networks, a large amount of works have proposed to model sequential user behaviors by using various neural network, such as self-attention network [6,10], graph neural network [13]. Moreover, some approaches [17] focus on incorporating the structure attribute information about items into the sequential recommendation models. Despite the success of these sequential recommendation models, they are highly dependent on the recommendation data due to the limitation of ID-based item representations.

Pre-training for Recommendation. Pre-training aims to learn useful representation or model architecture from large scale data, which is beneficial for the specific downstream tasks. In recommender systems, some works usually design self-supervised pre-training task to capture information from user behaviors [7,10]. Besides, some works try to incorporate structure attribute information about items [19] or users [14] into recommendation model through specific pre-training task. Different from these studies, this approach focuses on utilizing text semantic knowledge to develop ID-agonistic user behavior models through single-domain pre-training. Based on this work, UniSRec [5] further proposes a more general sequence representation learning approach for multiple-domain pre-training, which aims to learn universal representations across different recommendation domains or platforms.

6 Conclusion

In this paper, we propose the ID-agnostic user behavior modeling approach for sequential recommendation, named IDA-SR. Different from the existing sequential recommendation methods that are limited by the ID-based item representations, the proposed IDA-SR adopts the ID-agnostic item representations based on item texts to help user behavior modeling in a direct and natural way. To bridge the gap between text semantics and sequential user behaviors, the proposed IDA-SR conducts a pre-training architecture over item text representations on the sequential user behaviors. Experimental results have shown the effectiveness of the proposed approach by comparing it with several competitive baselines, especially when training data is limited.

Acknowledgements. This work was partially supported by the National Natural Science Foundation of China under Grant No. 61872369 and 61832017, Beijing Outstanding Young Scientist Program under Grant No. BJJWZYJH012019100020098.

References

1. Chen, C., Zhang, M., Liu, Y., Ma, S.: Neural attentional rating regression with review-level explanations. In: WWW, pp. 1583–1592 (2018)
2. Devlin, J., Chang, M., Lee, K., Toutanova, K.: BERT: pre-training of deep bidirectional transformers for language understanding. In: NAACL, pp. 4171–4186 (2019)
3. Ding, H., Ma, Y., Deoras, A., Wang, Y., Wang, H.: Zero-shot recommender systems. arXiv preprint arXiv:2105.08318 (2021)
4. Hidasi, B., Karatzoglou, A., Baltrunas, L., Tikk, D.: Session-based recommendations with recurrent neural networks. In: ICLR. arXiv preprint arXiv:1511.06939 (2016)
5. Hou, Y., Mu, S., Zhao, W.X., Li, Y., Ding, B., Wen, J.: Towards universal sequence representation learning for recommender systems. In: KDD, pp. 585–593 (2022)
6. Kang, W., McAuley, J.J.: Self-attentive sequential recommendation. In: ICDM, pp. 197–206 (2018)
7. Liu, Z., Fan, Z., Wang, Y., Yu, P.S.: Augmenting sequential recommendation with pseudo-prior items via reversely pre-training transformer. In: SIGIR, pp. 1608–1612 (2021)
8. Ni, J., Li, J., McAuley, J.J.: Justifying recommendations using distantly-labeled reviews and fine-grained aspects. In: EMNLP, pp. 188–197 (2019)
9. Rendle, S., Freudenthaler, C., Schmidt-Thieme, L.: Factorizing personalized Markov chains for next-basket recommendation. In: WWW, pp. 811–820 (2010)
10. Sun, F., et al.: BERT4Rec: sequential recommendation with bidirectional encoder representations from transformer. In: CIKM, pp. 1441–1450 (2019)
11. Tang, J., Wang, K.: Personalized top-n sequential recommendation via convolutional sequence embedding. In: WSDM, pp. 565–573 (2018)
12. Vaswani, A., et al.: Attention is all you need. In: NIPS, pp. 5998–6008 (2017)
13. Wu, S., Tang, Y., Zhu, Y., Wang, L., Xie, X., Tan, T.: Session-based recommendation with graph neural networks. In: AAAI, pp. 346–353 (2019)
14. Xiao, C., et al.: UPRec: user-aware pre-training for recommender systems. arXiv preprint arXiv:2102.10989 (2021)
15. Yang, Z., Dai, Z., Yang, Y., Carbonell, J.G., Salakhutdinov, R., Le, Q.V.: XLNet: generalized autoregressive pretraining for language understanding. In: NeurIPS (2019)
16. Yu, W., Lin, X., Ge, J., Ou, W., Qin, Z.: Semi-supervised collaborative filtering by text-enhanced domain adaptation. In: SIGKDD, pp. 2136–2144 (2020)
17. Zhang, T., et al.: Feature-level deeper self-attention network for sequential recommendation. In: IJCAI, pp. 4320–4326 (2019)
18. Zhao, W.X., et al.: RecBole: towards a unified, comprehensive and efficient framework for recommendation algorithms. In: CIKM, pp. 4653–4664 (2021)
19. Zhou, K., et al.: S3-Rec: self-supervised learning for sequential recommendation with mutual information maximization. In: CIKM, pp. 1893–1902 (2020)

Enhance Performance of Ad-hoc Search via Prompt Learning

Shenghao Yang, Yiqun Liu(✉), Xiaohui Xie, Min Zhang, and Shaoping Ma

Department of Computer Science and Technology, Beijing National Research Center for Information Science and Technology, Tsinghua University, Beijing 100084, China
ysh21@mails.tsinghua.edu.cn, {yiqunliu,z-m,msp}@tsinghua.edu.cn,
xiexiaohui@mail.tsinghua.edu.cn

Abstract. Recently, pre-trained language models (PTM) have achieved great success on ad hoc search. However, the performance decline in low-resource scenarios demonstrates the capability of PTM has not been inspired fully. As a novel paradigm to apply PTM to downstream tasks, prompt learning is a feasible scheme to boost PTM's performance by aligning the pre-training task and downstream task. This paper investigates the effectiveness of the standard prompt learning paradigm on the ad hoc search task. Based on various PTMs, two types of prompts are tailored for the ad hoc search task. Overall experimental results on the MS Marco dataset show the credible better performance of our prompt learning method than fine-tuning based methods and another previous prompt learning based model. Experiments conducted in various resource scenarios show the stability of prompt learning. RoBERTa and T5 deliver better results compared to BM25 using 100 training queries utilizing prompt learning, while fine-tuning based methods need more data. Further analysis shows the significance of the uniformity of tasks' format and adding continuous tokens into training in our prompt learning method.

Keywords: Ad hoc search · Prompt learning · Pre-trained language model

1 Introduction

In recent years, large-scale pre-trained language models (PTM), e.g. BERT [1], have boosted numerous downstream tasks' performance. In practice, the "pre-train and fine-tuning" paradigm is widely adopted. For ad hoc search, especially in the second stage (i.e. re-ranking), the query and document are usually concatenated and fed into the encoder of PTM in fine-tuning [2]. The credible relevance score is obtained utilizing the efficient self-attention mechanism. However, it has been pointed out that this paradigm cannot fully inspire the capabilities of PTM. The main reason is the gap in the form of pre-training and downstream tasks [3]. We argue that this problem also limited the performance of PTM on the ad hoc search task. Especially in the low-resource scenarios, the performance decline obviously [4].

Y. Chang and X. Zhu (Eds.): CCIR 2022, LNCS 13819, pp. 28–39, 2023.
https://doi.org/10.1007/978-3-031-24755-2_3

To tackle this problem, a potentially feasible method is adopting a novel paradigm, namely prompt learning, on the ad hoc search task instead of the "pre-train and fine-tuning" paradigm. This new paradigm is describe as "pre-train, prompt, and predict" [3]. As distinguished from fine-tuning, it is called prompt-tuning. It boosts the performance of PTM by aligning the form of pre-training and fine-tuning. For example, the Masked language model (MLM) is a popular pre-training task and aims to accomplish a cloze style task. Intuitively, converting the form of the downstream task into cloze style can contribute to the performance. Utilizing the powerful natural language understanding ability of MLM, cloze style prompt-tuning is commonly adopted in many tasks and has achieved convincing performance [5–8].

Research on prompt learning for the ad hoc search task is not sufficient. MonoT5 [9] converts the re-ranking task into a text generation task and utilizes a Seq2Seq model (i.e. T5 [10]) to perform task. P^3 ranker [11] follows monoT5 and proposes a pre-fine-tuning scheme to boost performance. It first conducts prompt-tuning on a natural language inference (NLI) dataset MNLI [12] and further experiments on ad hoc search datasets.

Compared with existing research, we adopt the standard prompt learning (i.e. "pre-train, prompt, and predict" paradigm) without additional training data and training stage. We investigate the prompt-tuning performance of lightweight models (i.e. BERT and RoBERTa [13]). In the experiments based on a popular ad hoc search benchmark dataset (i.e. MS Marco [14]), RoBERTa with prompt-tuning achieves better performance compared to the BERT and RoBERTa with fine-tuning, especially in the low-resource setting. The experimental results demonstrate the effectiveness of prompt learning and it is more convincing since the uniformity of model architecture. We further apply our method to T5 and achieve further performance improvement.

In summary, the contributions of our work are as follows:

- We investigate the effectiveness of the standard prompt learning paradigm on the ad hoc search task, especially in low-resource scenarios.
- We tailor the prompt-tuning method for the ad hoc search task and summarize the performance of using various PTMs as the backbone and utilizing various prompt types.
- Experimental results on MS Marco demonstrate the effectiveness of our prompt learning method regardless in low-resource and full-resource scenarios. Further analysis demonstrates that aligning the task format and adding the continuous tokens contribute to our method's performance.

2 Related Work

2.1 Ad Hoc Search with PTM

Utilizing the outstanding natural language model ability of PTM, the performance of ad hoc search has gained significant improvements. In the retrieval stage of ad hoc search, differing from some sparse retrieval models, e.g. BM25 [15],

DPR [16] applies PTM to obtain the dense representations of query and document. The relevance score is the dot product between the representations of query and document. Then the approximate nearest neighbor (ANN) algorithm is adopted to perform retrieval. The following researches [17,18] mainly focus on the hard negative sampling strategy based on the framework of DPR. In the re-ranking stage of ad hoc search, [2] proposed to obtain the relevance score with a vanilla BERT attached with a MLP head. It concatenates query and candidate passage retrieved in the first stage with a special token (i.e. [SEP] token) as the input of BERT. This scheme is widely adopted in the following researches [19–21]. However, it has been witnessed an obviously performance decline when training data is insufficient [4]. In this paper, we focus on alleviating the performance decline in the re-ranking stage. We speculate the aforementioned scheme may not attain the best potential performance and we are concerned with a different scheme, called prompt learning.

2.2 Prompt Learning

Prompt learning has achieved remarkable development and it has been extensive study and applications in recent years. In GPT3 [6], tasks such as translation can be accomplished by adding contextual prompts. This scheme accomplishes the aligning of the pre-training task and downstream task and achieves performance boosting.

At the early development phase of prompt learning, prompts were generally designed manually [5,6]. However, subtle differences in the manual design will cause a significant impact on performance [7]. In that regard, many automatic approaches have been proposed to find suitable prompts. For prompt in natural language format, suitable template words are mined in the discrete space [22]. Further, some approaches [7,8,23,24] get rid of the natural language format and adopt the continuous format, i.e., not using a specific word but embedding. The advantage of using continuous prompt is that the embedding can participate in the training as a trainable parameter.

Prompt learning has been applied in various NLP tasks, yet the research on ad hoc search is not enough. P^3 ranker [11] first brought the concept of prompt learning into ad hoc search, it follows the prompt of monoT5 [9] and adds a pre-fine-tuning stage to warm up the training process with MNLI [12] dataset. MNLI is a sentence pair classification task that is similar to the ad hoc search task. Thus, the pre-fine-tuning stage contributes to the performance. Above works all conduct prompt learning with discrete prompts. In this paper, we follow the standard prompt learning scheme and conduct a comprehensive experiment on both discrete and continuous prompts.

3 Preliminary

3.1 Ad hoc Search

In this section, we will briefly introduce the basic knowledge of the ad hoc search. Generally, the ad hoc search task is performed in two stages. Given queries, candidate documents are obtained using a retrieval system (e.g. BM25 [15]) in the first stage. The second stage, called re-ranking, calculates the relevance score of the query to each candidate document using more sophisticated methods (e.g. learning to rank algorithms and neural ranking models).

In this paper, we focus on the re-ranking task. It aims to build a ranking function f_r which can measure the relevance score $s_{i,j}$ for each query q_i and all its candidate documents $d_{i,j}$. The optimal ranking function f_r^* is obtained by minimizing the ranking loss L with supervised learning, which can be formalized as:

$$f_r^* = argmin\ L(f_r; Q, D, Y), \tag{1}$$

where Q, D, Y is the set of queries, candidate documents, and relevance labels, respectively. In the experiments of this paper, the ranking loss is chosen as MarginRankingLoss:

$$L(f_r; Q, D, Y) = \sum_{q_i \in Q} \sum_{d_j, d_k \in D} max(0, -y_{i,j,k}(r_{i,j,k} + m)) \tag{2}$$

where $r_{i,j,k} = f_r(q_i, d_j) - f_r(q_i, d_k)$, and $f_r(q, d)$ represents the relevance score calculated by ranking function f_r. Given q_i, $\begin{cases} y_{i,j,k} = 1, d_{i,j} \succ d_{i,k} \\ y_{i,j,k} = -1, d_{i,k} \succ d_{i,j} \end{cases}$, $\succ$ represents more relevant, m is set to 1.

3.2 Prompt Learning

In this section, we will briefly introduce the basic concepts of prompt learning. We will take the sentiment analysis task as an example.

Template. Consider a sentence "[X], it is ___". This sentence with a slot is called the template in prompt learning. [X] is a placeholder for input text. We formally defined a template function f_t. It outputs x' after inserting the input x into the template, i.e. $x' = f_t(x)$. x' will be fed into PTM to obtain the probability of the token at the slot position.

Verbalizer. For the sentence above, if we define the word "good" corresponds to the positive label, the prediction probability of "good" at the slot position can be regarded as the positive probability. The word is called a label word. The process of mapping words to labels is called verbalizer and can be defined as a function f_v. Given a PTM, the probability of label word w_i is formalized as $P(s = w_i | x', \theta_m)$, where θ_m is the parameters of the PTM and s is the token at slot position. In summary, given the input x, we obtain the probability of the label y by Eq. 3.

$$P(y|x; \theta_m) = \frac{1}{N_y} \sum_{w_i \in f_v(y)} P(s = w_i | f_t(x); \theta_m), \tag{3}$$

where $f_v(y)$ mapping label y to label words and N_y is the number of label words of label y.

4 Methodology

In this section, we will present our method to apply prompt learning to the re-ranking task. For this task, the input x contains two parts, i.e. query and document. Thus, two placeholders need to be set in the template, namely [q] and [d] corresponding to the query and document, respectively. The slot position is represented with a special token [mask]. For the verbalizer, we assign one label word per label. Specifically, the prompt tailored for the re-ranking task in various PTM settings is shown in Table 1. We adopt these prompts after a few trials. In other words, these prompts achieve better performance in the experiment.

Table 1. Templates and verbalizers tailored for re-ranking task

PTM	Template	Verbalizer (pos/neg)
BERT/RoBERTa	[q] and [d] are [mask]	Relevant/irrelevant
	[q] [soft]×3 [mask] [soft]×3 [d]	Yes/but
T5	Query: [q] Document: [d] Relevant: [mask]	True/false
	[soft] [q] [soft] [d] [soft] [mask]	

We summarize our method into two type: *hard prompt* and *soft prompt*. Specifically, *hard prompt* is designed entirely in natural language and *soft prompt* may contain continuous embedding. In the training of *hard prompt* based method, only the PTM's parameters need to be optimized. The label probability is predicted using Eq. 3. This process is shown as Fig. 1(a).

For *soft prompt*, the template contains [soft] placeholder as shown in Table 1. It will be initialized with word embedding and optimized as parameters in training. For BERT/RoBERTa, the word embedding is initialized randomly. For T5, the three [soft] placeholder will be initialized by "Query:", "Document:" and "Relevant:", respectively. Specifically, the raw token is fed to the word embedding layer of the PTM, while the soft token is fed to a custom soft embedding layer. This scheme is followed by [7]. The outputs of both embedding layers are then combined and fed to the following layers.

In the verbalizer process of *soft prompt*, we follow the setting in [24]. Unlike utilizing the probability of label words, the last hidden state of the [mask] token is fed into a fully-connected (FC) layer to obtain label probability. The weight of the FC layer is initialized with label words' embedding. The framework of *soft prompt* is shown in Fig. 1(b). We set two separate optimizers for the soft embedding layer and the FC layer.

In the case of *soft prompt*, Eq. 3 will be take the form of Eq. 4:

$$P(y|x;\theta_m,\theta_v,\theta_t) = \frac{exp\theta_v^y f_m(f_t(x);\theta_m,\theta_t)}{\sum\limits_{i\in C} exp\theta_v^{y_i} f_m(f_t(x);\theta_m,\theta_t)}, \tag{4}$$

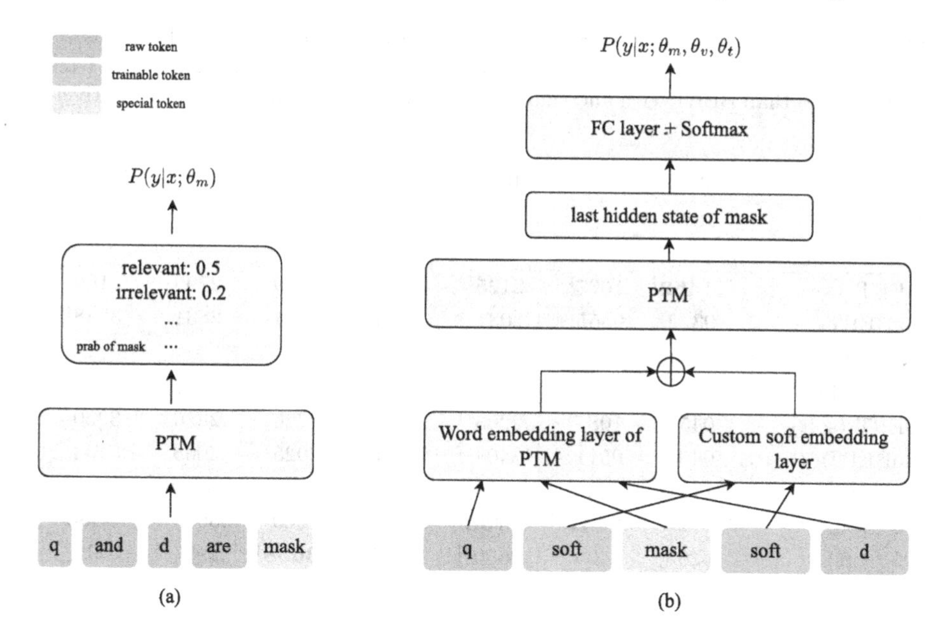

Fig. 1. The prompt learning framework tailored for the re-ranking task: (a) *hard prompt*, (b) *soft prompt*. Note that this figure shows the template and verbalizer of BERT/RoBERTa. The framework with T5 is broadly consistent.

where $f_m(\cdot)$ is the last hidden state of [mask] token output by PTM, θ_v^y is the label word's embedding corresponding to label y, and C is the number of labels. Compared to *hard prompt*, *soft prompt* put template and verbalizer into training parameters, which benefit to alleviate the bottleneck of manual design.

Finally, to obtain the relevance scores of the query-document pair in the input x, we subtract the probability of negative label from the probability of positive label, i.e. $f_r(x) = P(y = 1|x;\theta) - P(y = 0|x;\theta)$, where θ represents all possible parameters. $f_r(x)$ can be involved in the calculation of the ranking loss as $f_r(q, d)$ in Eq. 2 and optimized by Eq. 1.

5 Experiments

In this section, we will introduce the experimental settings and results in detail.

5.1 Dataset and Metric

We experiment on the MS MARCO passage ranking dataset [14]. It contains approximately 530 k queries in the training set and 6,980 queries in the development set. Each query has one relevant document on average.

To investigate the performance of prompt learning in various resource scenarios, we sample queries to construct the training sets. Specifically, the training

Table 2. Overall performance of our prompt learning method on MS MARCO passage ranking dataset. †‡ represent significant performance improvement using pairwise t-test with $p < 0.05$ than BERT with fine-tuning and RoBERTa with fine-tuning, respectively.

Model	50	100	500	1 k	5 k	10 k	530 k
BM25	.1874						
Fine-tuning based models							
BERT	**.1462**‡	.1677‡	.2125‡	.2309	.2760	.2792	.3109
RoBERTa	.0322	.0855	.2031	.2524†	.2874†	.2831	.3248†
Prompt-tuning based models							
BERT (hard)	.0316	.0372	.1514	.2208	.2733	.2776	.3038
BERT (soft)	.0454‡	.1087‡	.2203†‡	.2283	.2726	.2826	.3060
RoBERTa (hard)	.0408‡	.0511	.2404†‡	.2581†	.2925†	.2968†‡	.3133†
RoBERTa (soft)	.1121‡	.1936†‡	.2336†‡	.2617†‡	.2932†‡	**.3028**†‡	.3182†
T5 (hard)	.1121‡	**.2142**†‡	**.2730**†‡	**.2833**†‡	.3029†‡	.3000†‡	**.3334**†‡
T5 (soft)	.0269	.1400‡	.2650†‡	.2763†‡	**.3038**†‡	.2997†‡	.3219†

sets contain 50/100/500/1 k/5 k/10 k/530 k (all) queries, respectively. We don't construct training sets with too few queries (e.g. 5/10 queries) considering the performance will depend heavily on the sample queries and more randomness. To conduct pairwise training, each query in the training set is paired with a relevant document and an irrelevant document. The irrelevant document is sampled from hard negative documents retrieved by BM25. For each training set, we sample queries from the development set as the validation set. In the 50/100/500 scenarios, we sample equally queries in the validation set, and we all sample 500 queries in 1 k/5 k/10 k/530 k scenarios.

When we evaluate our method on the validation set and development set, we first retrieved the top 100 documents for each query using BM25 and then re-rank the candidate documents. The evaluation metric is MRR@10.

5.2 Experimental Setup

We take three types of models as the baseline in our experiment. A lexical model, BM25. Two fine-tuning based models, BERT and RoBERTa with fine-tuning. A prompt-tuning based model, P^3 ranker [11].

Our experimental framework is implemented based on Openprompt [25] and transformer [26] library. We chose BERT/RoBERTa/T5 as the backbone of our models. We adopt the base version of these models. The prompts tailored for these models have been presented in Sect. 4.

In the experiment, we train our models for 100 epochs and set the batch size to 10 and gradient accumulation steps to 2. The max length of the input is set to 256. We use ADAM [22] with the initial learning rate set to 3e-5, $\beta1$=0.9, $\beta2$=0.999, epsilon=1e-8, L2 weight decay of 0.01, learning rate warmup over the first 10% steps, and linear decay of the learning rate.

Table 3. Performance comparison between our prompt learning method and P[3]ranker [11] on MS MARCO Passage Ranking.

Model	50	1 k	530 k
P3 ranker	0.0949	0.2027	0.3311
BERT (hard)	0.0316	0.2208	0.3038
BERT (soft)	0.0454	0.2283	0.3060
RoBERTa (hard)	0.0408	0.2581	0.3133
RoBERTa (soft)	0.1120	0.2617	0.3182
T5 (hard)	**0.1121**	**0.2833**	**0.3334**
T5 (soft)	0.0269	0.2763	0.3219

5.3 Result and Analysis

The overall performance of our prompt learning method is shown in Table 2. From the results, it is clear that the prompt-tuning models outperform fine-tuning and lexical methods except in the 50 queries scenario. We will analyze this exception in 5.3.1. T5 and RoBERTa with prompt-tuning achieve better results than BM25 using 100 training queries while fine-tuning method needs more training queries. As shown in Table 3, compared to another prompt learning based method, i.e. P[3] ranker, superior results are seen when using our method.

5.3.1 Prompt-Tuning vs Fine-Tuning

From experimental results, we find that T5 with prompt-tuning can achieve overall better performance than fine-tuning methods. This is in line with our expectations considering T5 has more parameters and a more sophisticated pre-training scheme. One exceptional case is found in the 50 queries setting. The BERT with fine-tuning performs well in that case.

We will explain the experimental results of four methods with the same model architecture from the perspective of format aligning. Table 4 compares the format of BERT and RoBERTa in various tasks. Combined with the experimental results, we have several findings:

- BERT with prompt-tuning underperforms BERT with fine-tuning. It can be seen in Fig. 2(b). We speculate the reason is that BERT retains the next sentence prediction (NSP) task in pre-training. Considering the NSP task and re-ranking task are both text pair classification tasks and take the same input format. That can be seen as a uniformity between the pre-training task and downstream task and contribute to the performance. This may also be the reason why BERT with fine-tuning performs well in the 50 queries setting.
- However, The good trend of BERT with fine-tuning is not maintained when the number of training queries increases while RoBERTa's performance increases rapidly with more training queries. It can be shown in Fig. 2(a). The result is reasonable since RoBERTa is an optimal version of BERT.

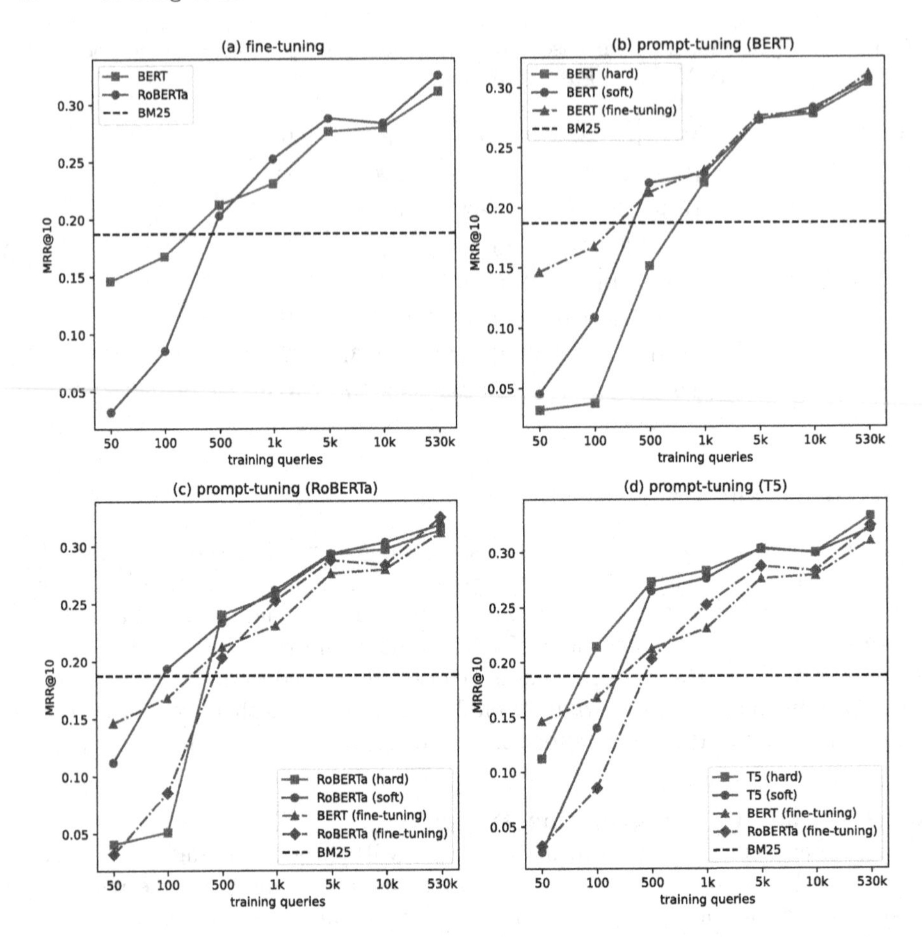

Fig. 2. The performance with various PTM backbones and various prompts in each resource scenario.

- RoBERTa with prompt-tuning gives clearly better results than BERT with prompt-tuning. This demonstrates that RoBERTa is more appropriate for the cloze style prompt learning method since RoBERTa only retains the MLM task in pre-training and conducts more training iterations. Superior results are seen for RoBERTa with prompt-tuning compared with two fine-tuning methods. It can be seen in Fig. 2(c). Specifically, prompt-tuning boosts the performance of RoBERTa in very low-resource scenarios and this demonstrates the advantage of prompt learning to inspire the capacity of PTM.

5.3.2 Soft Prompt vs Hard Prompt

In our prompt learning method, each PTM includes two types of prompts (detail in Sect. 4). For BERT/RoBERTa, they deliver significantly better results due to soft prompt in low-resource scenarios (i.e. 50/100 setting). In particular,

Table 4. Comparison of the format in different types of tasks using BERT and RoBERTa.

Task	PTM	Task type	Input format
MLM	BERT, RoBERTa	Pre-training	$[t_1]$ $[t_2]$ ... $[mask]$ ... $[t_n]$
NSP	BERT	Pre-training	$[s_1]$ $[sep]$ $[s_2]$
Re-ranking	BERT, RoBERTa	Fine-tuning	$[q]$ $[sep]$ $[d]$
		Prompt-tuning (hard)	$[q]$ *and* $[d]$ *are* $[mask]$
		Prompt-tuning (soft)	$[q]$ $[soft]$ $[mask]$ $[soft]$ $[d]$

RoBERTa can give a better result than BM25 with soft prompt-tuning using 100 training queries. It can be seen in Fig. 2(c). In the setting with more training queries, the two types of prompts show similar performance. However, the advantage of the soft prompt doesn't appear for T5. The performance even declines in several scenarios and the trend intensifies when using fewer training queries. We speculate that it might be due to the soft prompt we design for T5 is not potentially optimal. Even so, T5 with hard prompt-tuning has achieved superior results compared with other methods and we believe that a more sophisticated soft prompt for T5 will further boost its performance.

5.4 Case Study

In this section, we will investigate some cases in which the prompt-tuning based model significantly improves the performance compared to the fine-tuning based model. Specifically, we choose RoBERTa with hard prompt-tuning and RoBERTa with fine-tuning in the 10 k queries setting for the comparison. Intuitively, prompt-tuning is a token classification task that focuses on the semantic match at the token level. While as a text pair classification task, fine-tuning can capture the text level relevance signal but it may neglect the lexical match.

In the case shown in Table 5, fine-tuning based model ranks an irrelevant document at the first position since it is a guide for one job in texas. However, the job in query is a broker rather than a notary. Fine-tuning based model only considers the coarse-grained semantic match while neglecting the keyword "broker". Prompt-tuning based model can capture the keyword information in the relevant document and ranks it at the first position while fine-tuning based model ranks it at the 13th position.

Table 5. Case study on query 393696. Document 759874 (second row) is a relevant document and document 1455664 (third row) is an irrelevant document for this query, respectively.

Query: in texas what requirements are needed to apply to become a broker
The Texas State Securities Board has no established educational requirements for becoming a stockbroker, but many broker-dealer firms will require that applicants for sponsorship hold a bachelor's degree. Additionally, if you choose to pursue professional designations during your career, a bachelor's degree, at minimum, is usually required. **Rank result:** RoBERTa with hard prompt-tuning (1), RoBERTa with fine-tuning (15).
Here is a step-by-step guide to become a notary in Texas. Notaries.com handles all of these requirements through our easy application, so rest assured you have everything you need to become a notary! In order to become a notary in the State of Texas, you must: Be at least 18 years old. **Rank result:** RoBERTa with hard prompt-tuning (60), RoBERTa with fine-tuning (1)

6 Conclusion

In this paper, we investigate the performance of prompt learning in the ad hoc search task. We first introduce the formulaic representation of the ad hoc search task and prompt learning. Then we design two types of prompts, i.e. hard and soft prompts, for three PTM, i.e. BERT, RoBERTa and T5. With a prompt learning framework tailored for the ad hoc search task, we conduct experiments on MS Marco dataset. The experimental results in full-resource and low-resource scenarios show a convincing performance of our prompt learning method compared with baseline methods. We further analyze the experimental results and show that aligning the task format and utilizing soft prompt can contribute to better results in our method. In the case study, we note that prompt-tuning and fine-tuning have different relevance signal preferences. This inspires us to further analyze how prompt learning methods enhance the performance of the ad hoc search in the future.

Acknowledgments. This work is supported by the Natural Science Foundation of China (Grant No. 61732008) and Tsinghua University Guoqiang Research Institute.

References

1. Devlin, J., Chang, M.-W., Lee, K., Toutanova, K.: BERT: pre-training of deep bidirectional transformers for language understanding. arXiv preprint arXiv:1810.04805 (2018)
2. Nogueira, R., Cho, K.: Passage re-ranking with BERT. arXiv preprint arXiv:1901.04085 (2019)
3. Liu, P., Yuan, W., Fu, J., Jiang, Z., Hayashi, H., Neubig, G.: Pre-train, prompt, and predict: a systematic survey of prompting methods in natural language processing. arXiv preprint arXiv:2107.13586 (2021)

4. Zhang, X., Yates, A., Lin, J.: A little bit is worse than none: ranking with limited training data. In: Proceedings of SustaiNLP: Workshop on Simple and Efficient Natural Language Processing, pp. 107–112 (2020)
5. Petroni, F., et al.: Language models as knowledge bases? arXiv preprint arXiv:1909.01066 (2019)
6. Brown, T., et al.: Language models are few-shot learners. Adv. Neural. Inf. Process. Syst. **33**, 1877–1901 (2020)
7. Liu, X., et al.: GPT understands, too. arXiv preprint arXiv:2103.10385 (2021)
8. Han, X., Zhao, W., Ding, N., Liu, Z., Sun, M.: PTR: prompt tuning with rules for text classification. arXiv preprint arXiv:2105.11259 (2021)
9. Nogueira, R., Jiang, Z., Lin, J.: Document ranking with a pretrained sequence-to-sequence model. arXiv preprint arXiv:2003.06713 (2020)
10. Raffel, C., et al.: Exploring the limits of transfer learning with a unified text-to-text transformer. J. Mach. Learn. Res. **21**(140), 1–67 (2020)
11. Hu, X., Yu, S., Xiong, C., Liu, Z., Liu, Z., Yu, G.: P ^{3} ranker: mitigating the gaps between pre-training and ranking fine-tuning with prompt-based learning and pre-finetuning. arXiv preprint arXiv:2205.01886 (2022)
12. Williams, A., Nangia, N., Bowman, S.R.: A broad-coverage challenge corpus for sentence understanding through inference. arXiv preprint arXiv:1704.05426 (2017)
13. Liu, Y., et al.: RoBERTa: a robustly optimized BERT pretraining approach. arXiv preprint arXiv:1907.11692 (2019)
14. Nguyen, T., et al.: MS MARCO: a human generated machine reading comprehension dataset. In: CoCo@ NIPS (2016)
15. Robertson, S.E., Walker, S., Jones, S., Hancock-Beaulieu, M.M., Gatford, M., et al.: Okapi at TREC-3. NIST Spec. Publ. **109**, 109 (1995)
16. Karpukhin, V., et al.: Dense passage retrieval for open-domain question answering. arXiv preprint arXiv:2004.04906 (2020)
17. Xiong, L., et al.: Approximate nearest neighbor negative contrastive learning for dense text retrieval. arXiv preprint arXiv:2007.00808 (2020)
18. Qu, Y., et al.: RocketQA: an optimized training approach to dense passage retrieval for open-domain question answering. arXiv preprint arXiv:2010.08191 (2020)
19. Li, C., Yates, A., MacAvaney, S., He, B., Sun, Y.: PARADE: passage representation aggregation for document reranking. arXiv preprint arXiv:2008.09093 (2020)
20. Dai, Z., Callan, J.: Deeper text understanding for IR with contextual neural language modeling. In: Proceedings of the 42nd International ACM SIGIR Conference on Research and Development in Information Retrieval, pp. 985–988 (2019)
21. Nogueira, R., Yang, W., Cho, K., Lin, J.: Multi-stage document ranking with BERT. arXiv preprint arXiv:1910.14424 (2019)
22. Shin, T., Razeghi, Y., Logan IV, R.L., Wallace, E., Singh, S.: Autoprompt: eliciting knowledge from language models with automatically generated prompts. arXiv preprint arXiv:2010.15980 (2020)
23. Li, X.L., Liang, P.: Prefix-tuning: optimizing continuous prompts for generation. arXiv preprint arXiv:2101.00190 (2021)
24. Hambardzumyan, K., Khachatrian, H., May, J.: WARP: word-level adversarial reprogramming. arXiv preprint arXiv:2101.00121 (2021)
25. Ding, N., et al.: OpenPrompt: an open-source framework for prompt-learning. arXiv preprint arXiv:2111.01998 (2021)
26. Wolf, T., et al.: HuggingFace's transformers: state-of-the-art natural language processing. arXiv preprint arXiv:1910.03771 (2019)

Syntax-Aware Transformer for Sentence Classification

Jiajun Shan(✉), Zhiqiang Zhang(✉), Yuwei Zeng, Yuyan Ying, Haiyan Wu, Haiyu Song, Yanhong Chen, and Shengchun Deng

School of Information Management and Artificial Intelligence, Zhejiang University of Finance and Economics, Hangzhou 310020, Zhejiang, China
{shanjj,zqzhang,zyuwei,wuhy2020}@zufe.edu.cn, dsc@hit.edu.cn

Abstract. Sentence classification is a significant task in natural language processing (NLP) and is applied in many fields. The syntactic and semantic properties of words and phrases often determine the success of sentence classification. Previous approaches based on sequential modeling mainly ignored the explicit syntactic structures in a sentence. In this paper, we propose a **S**yntax-**A**ware **Trans**former (SA-Trans), which integrates syntactic information in the transformer and obtains sentence embeddings by combining syntactic and semantic information. We evaluate our SA-Trans on four benchmark classification datasets (i.e., AG'News, DBpedia, ARP, ARF), and the experimental results manifest that our SA-Trans model achieves competitive performance compared to the baseline models. Finally, the case study further demonstrates the importance of syntactic information for the classification task.

Keywords: Syntax · Transformer · Sentence classification

1 Introduction

Sentence classification, annotating pre-defined categories into free sentences, is an essential and significant task in many areas of natural language processing (NLP) applications. With the widespread availability of the Internet, numerous applications for sentence classification have emerged, for instance, sentiment analysis of customer comments and topic labeling based on news. In contrast to document-level classification, an individual sentence contains only a limited amount of information, hence, it is crucial to construct effective feature representations.

There is great progress in deep learning methods for sentence classification [1–3], most of these models (e.g., CNN, LSTM, GRU) based on sequential information have an excellent performance. Liu et al. [4] proposed three models of sharing information with RNN to capture the semantics of input text. Kim [5] proposed a convolutional neural network called TextCNN through learning task-specific vectors to improve classification performance. Nevertheless, it is tough

Y. Chang and X. Zhu (Eds.): CCIR 2022, LNCS 13819, pp. 40–50, 2023.
https://doi.org/10.1007/978-3-031-24755-2_4

for sequential models to capture the syntactic information that is vital for understanding the meaning of a sentence since these models lack any explicit modeling of syntax or any hierarchical structure of language [6]. This observation has raised interest in using deep learning approaches to study latent structure induction. From a practical perspective, it is also necessary to apply these tree-based structures to language modeling. More recently, there has been significant interest in exploiting syntactic knowledge. Roth and Lapata [7] embed the path of a dependency tree on the Neural Semantic Role Labeling task. Marcheggiani and Titov [8] apply Graph Convolutional Networks(GCN) to directly learn the dependency graph. These methods concentrate on exploiting dependency structures rather than constituency structures. Zhang et al. [9] propose a scheme to encode the syntax parse tree of sentences into a learnable distributed representation by encoding the path in the constituency tree corresponding to the word. Wu et al. [10] propose a modularized syntactic neural network to model constituency trees by utilizing the syntax category labels and context information. These methods enhance CNNs or RNNs by incorporating syntactic information. However, Transformer has achieved better performance than CNN and RNN on many NLP tasks. Previous researches show that although Transformer can learn syntactic knowledge purely by seeing examples, explicitly feeding this knowledge to Transformer is also beneficial [11].

In this work, we introduce a syntax-aware neural network to learn syntactic structure in the constituency tree better. We embed the latent tree structure through a recursive approach to extract the syntactic structure information of syntax trees. Meanwhile, we combine with the semantic information generated by Transformer for sentence classification. Our experiments on four benchmark datasets of sentence classification show that our proposed syntax-aware Transformer (SA-Trans) indeed obtains better performance. The major contributions of this work can be summarized as follows:

- We introduce a recursive-based method for embedding syntactic constituency information.
- A syntax-aware Transformer (SA-Trans) is proposed to combine syntactic and semantic features for sentence classification.
- Experimental results on four benchmark datasets show that SA-Trans significantly achieves competitive performance comparing with baselines.

2 Syntax-Aware Transformer

SA-Trans is composed of a syntactic subnetwork and a semantic subnetwork, whose structure is shown in Fig. 1. A merging layer then combines these two subnetworks to obtain the input of the final classification layer. The rest of the section describes the two subnetworks and the merging layer.

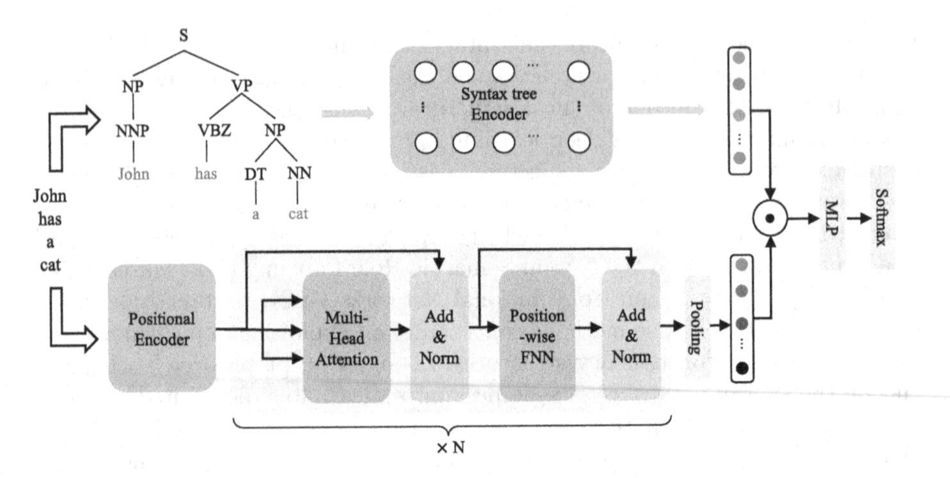

Fig. 1. The overall architecture of syntax-aware Transformer (SA-Trans). ***Syntactic Subnetwork*** embeds the syntactic structure of a syntax tree into a representation vector space. ***Semantic Subnetwork*** captures the semantic information of a sentence and induces semantic embeddings by a pooling operation. ***Merging Layer*** concatenates the syntactic vector and semantic vector and finally outputs the probabilities of each category.

2.1 Syntactic Subnetwork

In the syntactic subnetwork (see the top part of Fig. 1), the goal is to take the syntactic information of the input sentence and embed the hierarchically structured information of a syntax tree into a representation vector space. The Distributed Tree Embedding Layer (DTE) is the critical component of the syntactic subnetwork. DTE is based on a technique to embed the structural information of syntactic trees into dense, low-dimensional vectors of real numbers [12]. Given a sentence $S = [w_1, w_2, \ldots, w_L]$ comprising L-words, after parsing sentence into a syntax tree $\mathcal{T}$, DTE transforms syntax trees into low-dimensional vectors in a space $\mathbb{R}^d$ according to their subtrees. We use subtrees defined in Collins and Duffy [13]. See Fig. 2 for an example. Syntax trees $\mathcal{T}$ and their subtrees τ are recursively represented by $\mathcal{T} = (\gamma, [\tau_1, \ldots, \tau_m])$ where γ is the label representing the root of tree (like "S, NP, VP,..., NN") and $[\tau_1, \ldots, \tau_m]$ is a set of child trees. Note that we use $\mathcal{T} = (\gamma, [])$ with an empty list as trees to represent leaf nodes. The set $N(\mathcal{T})$ contains all the complete subtrees and valid subtrees of $\mathcal{T}$. Then the DTE is represented as the following embedding layer:

$$\boldsymbol{r} = W^{DTE} E(\mathcal{T}) \tag{1}$$

where $E(\mathcal{T}) = \mathbf{t}$ is a one-hot layer that maps subtrees to vectors in an n-dimension space $\mathbb{R}^n$ where n is the huge number of different subtrees (Conceptually we start with enumerating all subtrees that occur in the training data) and $W^{DTE}_{d\times n}$ is a matrix that transforms the large space of subtrees $\mathbb{R}^n$ into a

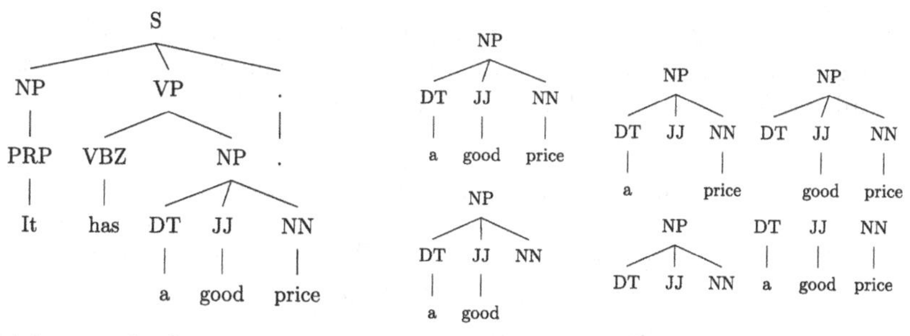

(a) An example of constituency tree.

(b) An example of complete subtrees and valid subtrees

Fig. 2. (a) An example of constituency tree. (b) The subtree of the NP node which covering *a good price*. Each tree in (a) contains all these subtrees, as well as many subtrees. Subtrees are mapped into an n-dimension representation space.

smaller space $\mathbb{R}^d$. Due to the large number of subtrees, this method is computationally expensive. We will consider reducing the computational complexity in our future work.

2.2 Semantic Subnetwork

The vanilla Transformer is an end-to-end neural network that has proven effective in many natural language processing tasks. We use a Transformer architecture similar to the setup in Vaswani et al. [14]. The transformer encodes the input sentence with a stack of N layers. Each layer has two sub-layers: one with multi-head self-attention and the other with a fully connected feed-forward neural network. At the input layer, a sentence of words $(w_1, w_2, \ldots, w_L)$ are embedded as $\mathbf{H}^0 = (\mathbf{w}_1, \mathbf{w}_2, \ldots, \mathbf{w}_L) \in \mathbb{R}^{L\times D}$, where D is the hidden size of model. A positional encoder then infuse position information into Transformer. After N layers multi-head self-attention and full connected feed-forward, we obtain the sequence embedding $\mathbf{H}^N = (\mathbf{w}_1^N, \mathbf{w}_2^N, \ldots, \mathbf{w}_L^N)$. In the end, a pooling operation employed for the last layer output to induce the semantic vector.

2.3 Merging Layer

The merging layer is composed of a concatenate layer with a ReLu activation function that concatenates the syntactic vector and the semantic vector. Eventually, a softmax layer calculates the probability of each category by output from Multilayer Perceptron (MLP).

3 Experiments

3.1 Datasets

To validate the performance of our SA-Trans, we apply four datasets for sentence classification from Wu et al. [10], which should be sensitive to syntactic information, including AG's News, DBpedia, Amazon Review Polarity (ARP), Amazon Review Full (ARF), where each sentence is annotated with one label. The detailed information of datasets are shown in Table 1.

Table 1. Statistics of evaluation datasets.

Datasets	Train	Dev	Test	Class
AG's News	108K	12K	7.6K	4
DBpedia	504K	56K	70K	14
ARP	400K	80K	80K	2
ARF	500K	100K	100K	5

3.2 Experimental Settings

This section describes the general experimental set-up of our experiments. For the Syntactic Subnetwork, each sentence's constituency tree is obtained by using Stanford's CoreNLP[1] probabilistic context-free grammar parser. Distributed trees are represented in a space $\mathbb{R}^d$ with $d = 300$. For the Transformer, as in prior work [14], we use dropout of 0.1, D of 512, d_{ff} of 2048, N_{layers} of 6 and the number of attention head is 8. The whole architecture is trained with Adam optimization method [15] in mini-batches at the size of 128 with a learning rate of 2e−4. We randomly initialized the word vectors. Note that we do not include any training tricks or pre-trained word embeddings here to avoid distractions and make the encoder structure affects the most. The result reports the average accuracy from 5 runs to control potential non-determinism associated with deep neural models [16]. All models, including baselines, are trained with a GPU (NVIDIA GeForce GTX 2080Ti).

3.3 Baseline Models

This section describes the baseline models, which are compared with our SA-Trans model.

- **TextCNN.** Kim [17] proposes an unbiased convolutional neural network model that attempts to use CNN for text classification, called TextCNN. We follow his settings: filter windows of 3, 4, 5 with 100 feature maps each, ReLU [18] activation function.
- **BiLSTM.** BiLSTM is proposed by Schuster and Paliwal [19] to access the preceding and succeeding semantics by combining the forward hidden layer and the backward hidden layer [20]. We use the memory dimension of BiLSTM of 32 and 2 hidden layers.

[1] https://stanfordnlp.github.io/CoreNLP.

- **BiLSTM+Attn.** Attention mechanism has been proved to improve the performance of text classification. Zhou et al. [21] applied the attention mechanism to the Cross-Lingual Sentiment Classification problem. In our experiment, we use BiLSTM+Attn with the same settings as BiLSTM.
- **KPCNN.** KPCNN [22] is a jointly model which can learn the coalesced embedding by using Convolutional Neural Network. This method leverages explicit knowledge and generates the implicit representation of the short text.
- **CCRCNN.** We also compared our work with a recently proposed method [23]. This method is based on recurrent convolutional neural network to capture context-relevant concept features.
- **Transfomer.** Transformer is a remarkable end-to-end model that has been widely applied in various domains [24]. In the sentence classification task, only the encoder of Transformer is used to model sentence embedding. The settings of the encoder we use are the same as in Sect. 3.2.
- **BERT-base.** BERT [25] is a pre-trained language representation model. BERT-base, used in the uncased setting with the pre-trained English model. We use the final hidden vector of the special *[CLS]* token for classification.

3.4 Results and Discussion

Table 2. The overall result on four benchmark datasets. The results without ⋆ of previous models are from our implementation since the dataset is different from theirs. ⋆ means the results are the results reported on their papers. - means not reported.

Model	AG's News	DBpedia	ARP	ARF
TextCNN	88.29	96.75	84.96	46.03
BiLSTM	88.17	96.95	**87.65**	49.15
BiLSTM+Attn	87.96	97.00	86.87	48.03
KPCNN [22]	88.36⋆	–	–	–
CCRCNN [23]	88.9⋆	–	–	–
BERT-base [26]	82.88⋆	97.11⋆	–	–
Transformer	88.54	96.50	85.58	47.66
Ours: SA-Trans	**90.20**	**97.81**	85.63	**49.29**

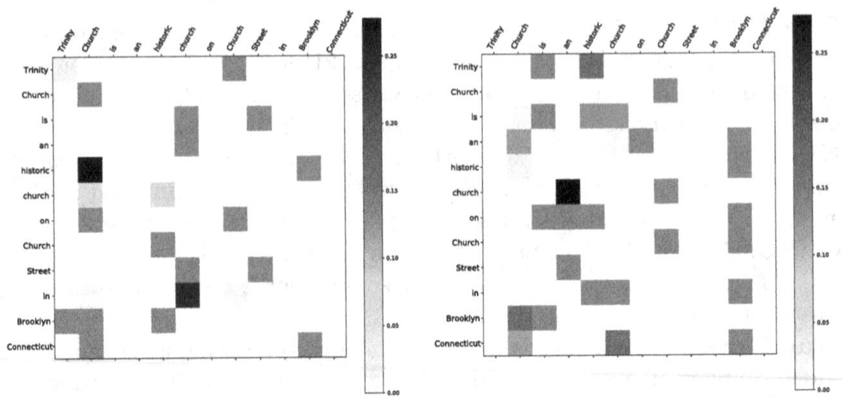

(a) The heat map of the SA-Trans with attention

(b) The heat map of the vanilla Transformer with attention.

Fig. 3. Heat map of attention of different models for the relations between words in sentence. In this case, the input sentence is *Trinity Church is an historic church on Church Street in Brooklyn Connecticut*, and the label is *Building*. The number of relations between words in subfigure (a) appears to be sparser than in subfigure (b) as the words (like "*is*", "*an*", "*on*", and "*in*") have less connection with other words. This phenomenon means that Transformer spends more effort on relationships that have a limited impact on the downstream classification task.

The overall performances are shown in Table 2, from which we can conclude several observations. The classification performance is evaluated on test datasets in terms of accuracy. Results from the experimental setting suggest that SA-Trans results are more robust than Transformer on AG's News, DBpedia, and ARP datasets, and the difference is statistically significant: 90.20 vs. 88.54 in AG's News, 97.81 vs. 96.50 in DBpedia, 85.63 vs. 85.58 in ARP, and 49.29 vs. 47.66 in ARF. The improvements of SA-Trans demonstrate that our model genuinely benefits from incorporating syntactic structure information. From Table 2, move forward a step, SA-Trans has been improved more significantly on AG's News (+1.66), DBpedia (+1.31) and ARF (+1.63) than on ARP (+0.05). The reason is that the sentences in AG's News and DBpedia are mostly high-quality and canonical declarative sentences, and their tree structures are less complicated compared to the reviews in Amazon datasets, which contain a certain amount of noise. Neither Transformer nor SA-Trans surpasses the sequential model BiLSTM on ARP. This is slightly surprising as Transformer should, in principle, be more competitive in terms of sentence modeling. We observe that the BiLSTM+Attention also performs weaker than BiLSTM on ARF and ARP. It seems that the noise in Amazon review datasets causes some interference with the attention mechanism.

SA-Trans achieves state-of-the-art performance on three datasets comparing baselines. This means SA-Trans obtain better sentence embedding with the aid of external syntactic information. We assume our superior mainly locate in attention, which will be interpreted in the following subsection.

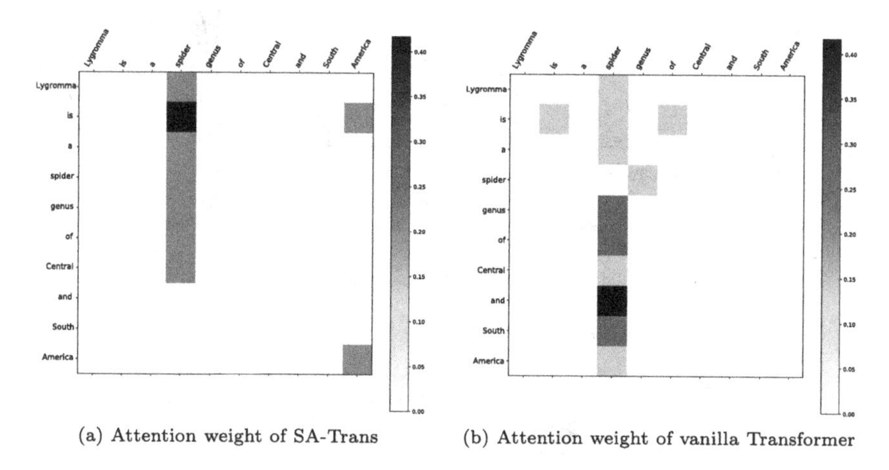

(a) Attention weight of SA-Trans

(b) Attention weight of vanilla Transformer

Fig. 4. Heat map of attention of different models for the relations between words in sentence. In this case, the input sentence is *Lygromma is a spider genus of Central and South America*, and the label is *Animal.* The number of relations between words in subfigure (a) appears to be sparser than in subfigure (b) as the words (like "*is*", "*a*", and "*of*") have less connection with other words. This phenomenon means that Transformer spends more effort on relationships that have a limited impact on the downstream classification task.

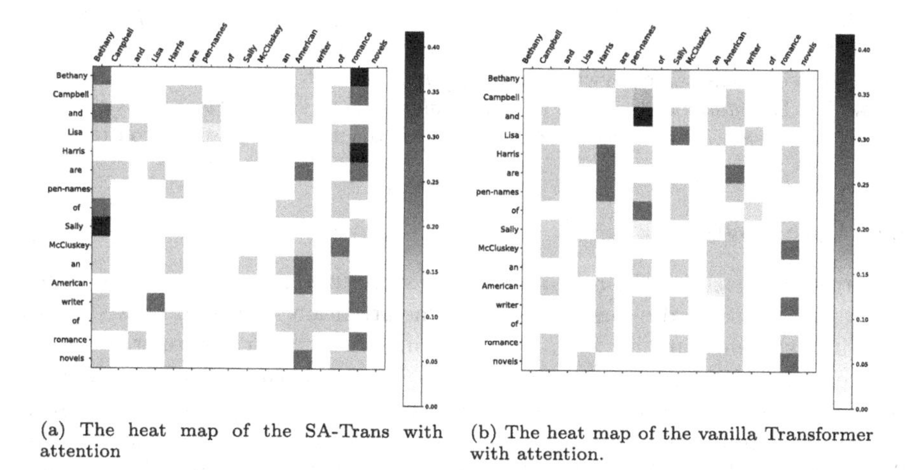

(a) The heat map of the SA-Trans with attention

(b) The heat map of the vanilla Transformer with attention.

Fig. 5. Heat map of attention of different models for the relations between words in sentence. In this case, the input sentence is *Bethany Campbell and Lisa Harris are pen-names of Sally McCluskey an American writer of romance novels*, and the label is *Artist.* The number of relations between words in subfigure (a) appears to be sparser than in subfigure (b), and subfigure (a) captures more relationships (the black blocks), which are beneficial for downstream classification tasks. This phenomenon implies that SA-Trans spends more effort on relationships that significantly impact the downstream classification task (Color figure online).

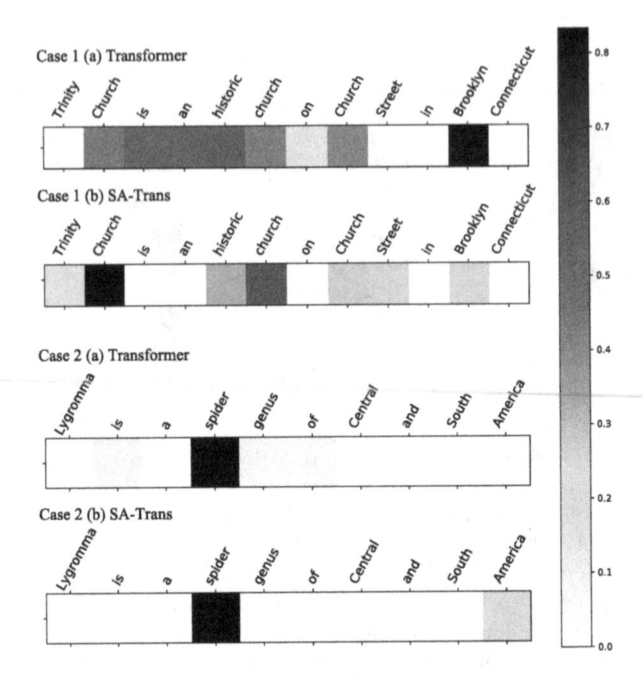

Fig. 6. Heat map of attention on each word in a sentence. In case 1, the input sentence is *Trinity Church is an historic church on Church Street in Brooklyn Connecticut*, the label is *Building*; In case 2, the input sentence is *Lygromma is a spider genus of Central and South America*, the label is *Animal.* Although both Transformer and SA-Trans focus on some same keywords (like "*Church* and "*Brooklyn*" in case 1, "*spider*" in case 2.), Transformer also looks at words (like "*is*", "*an*", and "*on*" in case 1, "*is*" and "*of*" in case 2.) that seem to be irrelevant for the sentence classification task.

3.5 Case Study

To further understand the role of syntactic structure in sentence classification, we present a case study. Firstly, We show the attention heat map between each word in a sentence in Figs. 3, 4, 5. Comparing with the Transformer, relations between words in the SA-Trans appear to be sparser because SA-Trans is more focused on the relations that are beneficial for the downstream task with the help of syntactic information. Furthermore, Fig. 6 illustrates the difference between Transformer and SA-Trans in focusing on each word. Although both Transformer and SA-Trans focus on some same keywords (like "*Church*" and "*Brooklyn*" in case 1, "*spider*" in case 2.), Transformer also looks at words (like "*is*", "*an*", and "*on*" in case 1, "*is*" and "*of*" in case 2.) that seem to be irrelevant for the sentence classification task. This indicates that syntactic structure information can guide the propagation of information in the network to allow model to attend to those words that have a more significant impact on correct classification and weaken the effect of words that are not relevant for classification results.

4 Conclusions

In this paper, a syntax-aware Transformer (SA-Trans) is proposed to combine syntactic structure information with semantic information for classification. We apply a recursive approach to embedding the latent tree structure and employ a Transformer model to capture the semantic information for sentence classification. We test the effectiveness of our model in the sentence classification task with four benchmark datasets. Our experimental results show that SA-Trans achieves powerful competitive performance, and syntactic structure has positive validity for sentence classification. In the future, we would like to investigate the performance of syntax structure by applying it directly to more Chinese language NLP tasks.

Acknowledgments. This work was partly supported by the National Natural Science Foundation of China under Grant (61972336, 62073284), and Zhejiang Provincial Natural Science Foundation of China under Grant (LY23F020001, LY22F020027, LY19F030008).

References

1. Krizhevsky, A., Sutskever, I., Hinton, G.E.: Imagenet classification with deep convolutional neural networks. In: Advanve in Neural Information Processing System, vol. 25, pp. 1097–1105 (2012)
2. Li, W., Qi, F., Tang, M., Yu, Z.: Bidirectional LSTM with self-attention mechanism and multi-channel features for sentiment classification. Neurocomputing **387**, 63–77 (2020)
3. Cho, K., et al.: Learning phrase representations using RNN encoder-decoder for statistical machine translation. In: EMNLP, pp. 1724–1734 (2014)
4. Liu, P., Qiu, X., Huang, X.: Recurrent neural network for text classification with multi-task learning. arXiv preprint arXiv:1605.05101 (2016)
5. Kim, Y.: Convolutional neural networks for sentence classification. In: Proceedings of the 2014 Conference on Empirical Methods in Natural Language Processing (EMNLP), pp. 1746–1751. Association for Computational Linguistics. Doha, Qatar (2014)
6. Bastings, J., Titov, I., Aziz, W., Marcheggiani, D., Sima'an, K.: Graph convolutional encoders for syntax-aware neural machine translation. In: EMNLP, pp. 1957–1967 (2017)
7. Roth, M., Lapata, M.: Neural semantic role labeling with dependency path embeddings. In: Proceedings of the 54th Annual Meeting of the Association for Computational Linguistics. ACL, pp. 1192–1202 (2016)
8. Marcheggiani, D., Titov, I.: Encoding sentences with graph convolutional networks for semantic role labeling. In: EMNLP, pp. 1506–1515 (2017)
9. Zhang, R., Hu, Z., Guo, H., Mao, Y.: Syntax encoding with application in authorship attribution. In: EMNLP, pp. 2742–2753 (2018)
10. Wu, H., Liu, Y., Shi, S.: Modularized syntactic neural networks for sentence classification. In: Proceedings of the 2020 Conference on Empirical Methods in Natural Language Processing (EMNLP), pp. 2786–2792. Association for Computational Linguistics, Online (2020)

11. Wang, J., Wei, K., Radfar, M., Zhang, W., Chung, C.: Encoding syntactic knowledge in transformer encoder for intent detection and slot filling. In: Proceedings of the AAAI Conference on Artificial Intelligence, vol. 35, no. 16, pp. 13943–13951 (2021)
12. Zanzotto, M.F., Dell'Arciprete, L.: Distributed tree kernels. In: Proceedings of the 29th International Conference on Machine Learning, ICML (2012)
13. Collins, M., Duffy, N.: New ranking algorithms for parsing and tagging: kernels over discrete structures, and the voted perceptron. In: Proceedings of the 40th Annual Meeting of the Association for Computational Linguistics, pp. 263–270 (2002)
14. Vaswani, A., et al.: Attention is all you need. In: Advances in Neural Information Processing Systems, pp. 5998–6008 (2017)
15. Kingma, D.P., Ba, J.: Adam: a method for stochastic optimization. In: Bengio, Y., LeCun, Y. (eds.) 3rd International Conference on Learning Representations, ICLR 2015, San Diego, CA, USA, 7–9 May 2015, Conference Track Proceedings (2015)
16. Reimers, N., Gurevych, I.: Reporting score distributions makes a difference: performance study of LSTM-networks for sequence tagging. In: EMNLP, pp. 338–348 (2017)
17. Kim, Y.: Convolutional neural networks for sentence classification. In: EMNLP, pp. 1746–1751 (2014)
18. Glorot, X., Bordes, A., Bengio, Y.: Deep sparse rectifier neural networks. In: Proceedings of the Fourteenth International Conference on Artificial Intelligence and Statistics, pp. 315–323 (2011). JMLR Workshop and Conference Proceedings
19. Schuster, M., Paliwal, K.K.: Bidirectional recurrent neural networks. IEEE Trans. Signal Process. **45**(11), 2673–2681 (1997)
20. Liu, G., Guo, J.: Bidirectional LSTM with attention mechanism and convolutional layer for text classification. Neurocomputing **337**, 325–338 (2019)
21. Zhou, X., Wan, X., Xiao, J.: Attention-based lstm network for cross-lingual sentiment classification. In: Proceedings of the 2016 Conference on Empirical Methods in Natural Language Processing, pp. 247–256 (2016)
22. Wang, J., Wang, Z., Zhang, D., Yan, J.: Combining knowledge with deep convolutional neural networks for short text classification. In: Proceedings of the Twenty-Sixth International Joint Conference on Artificial Intelligence, IJCAI-17, pp. 2915–2921 (2017)
23. Xu, J., Cai, Y.: Incorporating context-relevant knowledge into convolutional neural networks for short text classification. In: Proceedings of the AAAI Conference on Artificial Intelligence, vol. 33, no. 01, pp. 10067–10068 (2019)
24. Liu, D., Yang, K., Qu, Q., Lv, J.: Ancient modern Chinese translation with a new large training dataset. ACM Trans. Asian Low-Resour. Lang. Inf. Process. **19**(1), 1–13 (2019)
25. Devlin, J., Chang, M.-W., Lee, K., Toutanova, K.: Bert: Pre-training of deep bidirectional transformers for language understanding. arXiv preprint arXiv:1810.04805 (2018)
26. Zanzotto, F.M., Santilli, A., Ranaldi, L., Onorati, D., Tommasino, P., Fallucchi, F.: KERMIT: complementing transformer architectures with encoders of explicit syntactic interpretations. In: Proceedings of the 2020 Conference on Empirical Methods in Natural Language Processing (EMNLP), pp. 256–267. Association for Computational Linguistics, Online (2020)

Evaluation of Deep Reinforcement Learning Based Stock Trading

Yining Zhang(✉), Zherui Zhang, and Hongfei Yan

Peking University, Beijing, China
zhangyining@stu.pku.edu.cn

Abstract. Stock is one of the most important targets in investment. However, it is challenging to manually design a profitable strategy in the highly dynamic and complex stock market. Modern portfolio management usually employs quantitative trading, which utilizes computers to support decision-making or perform automated trading. Deep reinforcement learning (Deep RL) is an emerging machine learning technology that can solve multi-step optimal control problems. In this article, we propose a method to model multi-stock trading process according to reinforcement learning theory and implement our trading agents based on two popular actor-critic algorithms: A2C and PPO. We train and evaluate the agents on two datasets from 2010–2021 Chinese stock market multiple times. The experimental results show that both agents can achieve an annual return rate that outstrips the baseline by 8.8% and 16.8% on average on two datasets, respectively. Asset curve and asset distribution chart are plotted to prove that the policy the agent learned is reasonable. We also employ a track training strategy, which can further enhance the agent's performance by about 7.7% with little extra training time.

Keywords: Stock · Quantitative trading · Deep reinforcement learning · Actor-critic

1 Introduction

A company's stock is the product of a division of its ownership [14]. It serves as a way for companies to raise funds for development from investors. Since the concept appeared in 17th century, stock has become an irreplaceable component of the modern financial market. By the end of 2021, the ratio of Chinese stock market capitalization to GDP is 80.3% [4], while in the United States it has reached 201.9% [17]. However, the stock market is very volatile and unstable, which makes it extremely hard to manually design a profitable "silver-bullet" solution. Under this background, researchers have proposed a quantitative trading method that utilizes computers to build mathematical and statistical models based on massive historical data in order to find market signals that are likely to bring a high return. With the rapid development of deep learning in recent

Y. Chang and X. Zhu (Eds.): CCIR 2022, LNCS 13819, pp. 51–62, 2023.
https://doi.org/10.1007/978-3-031-24755-2_5

years, neural network models have been introduced into finance. As of today, the majority of research has been devoted to applying neural networks to predict future stock prices. These methods generally involve two steps: first, training a predicative model on historical dataset in a supervised manner; and second, using the prediction to derive a trading action so that the model can be tested by investment return rate or other metrics. The problem with these approaches is that the optimization objective is usually only designed for the first part, i.e., minimizing the prediction error, which is not the ultimate goal of investment. The second part, however, is likely to be some hand-crafted data-irrelevant rules, e.g., buy or sell a certain amount if the predicted stock price goes up or down over some threshold.

Meanwhile, the reinforcement learning (RL) approach, as another field in machine learning, can learn a direct mapping from market state to trading action to optimize a cumulative reward, which can be defined as the investment goal itself. Modern RL methods that utilize neural networks for function approximation are also called deep reinforcement learning. Though such works are relatively few in terms of publication numbers, the RL-based approach is considered a promising solution to quantitative trading for many reasons (see Sect. 3.4). A major problem we found in existing works is that most mainly focus on modeling aspects, but the experiment results are not shown in detail. Many of them only give the total return rate number or return curves without specifying whether it is the average or best result, and the hyperparameters used. Also, the validity of the model is barely verified through action playback to prove that the investment return is indeed earned with a reasonable trading strategy. According to [6,8], an RL model's performance may vary significantly between training and the policy that model learned may be sub-optimal and unacceptable, e.g., doing nothing at all in a game where most actions' reward is negative. These observations motivate us to do this fundamental research to draw a credible conclusion about the real performance of the RL model on stock trading via carrying out sufficient experiments and analysis.

In this paper, we first propose a method to model the stock trading process according to RL theory, specifically how to define the state, action, and reward functions, along with some assumptions about trading details, and then implement the RL agent based on two popular actor-critic algorithms: A2C and PPO[1]. We train and test our agent multiple times with different hyperparameters on two datasets from the 2010–2021 Chinese stock market, and report both the average performance and standard deviation. Considering the volatile nature of the stock market, we also employ an improving method similar to online learning and test it on two datasets. The result shows that both A2C and PPO agents can earn an annual return significantly higher than the baseline with a reasonable trading strategy, and the improving method can further enhance both agents' performance.

[1] Our code and dataset is available at Github: https://github.com/Altair-Alpha/DRL4StockTrading.

2 Related Works

In early times, quantitative trading was done by following certain trading rules proposed by investment experts. Today, there are still some works devoted to improving these classical methods, like [13], while others use them as baselines. With the rise of deep learning after 2010, researchers have introduced neural networks to build predicative models for financial products (stocks, futures, options, etc.). For example, [7] proposed a stock trading model based on a 2-D convolutional neural network (CNN). The representation of input data therefore became crucial and many works were dedicated to this topic, e.g., [2]. Meanwhile, some researchers tried to adopt a (deep) reinforcement learning approach to implement automated stock trading. The survey on RL for finance [5,11] and [12] demonstrates that existing works have used all three types of RL algorithms for finance, namely critic-only, actor-only, and actor-critic. Critic-only methods, mainly DQN and its variants, have been mostly explored, such as [3] and [9], as they are the early representatives of successful deep RL methods, while actor-critic methods are the least explored. There are also works that use an ensemble strategy to integrate multiple algorithms to achieve a better overall result, such as [16]. Similar to DL-based methods, state representation, as the input of an agent, was also a topic for RL. [15] utilized gated recurrent unit (GRU) for feature auto-extraction, and [1] introduced RCNN-based sentimental analysis as an extra feature. Unfortunately, we discovered that many previous works have some common deficiencies that makes reproducing and comparing their results difficult: (1) The implementation of RL algorithm (whether to use existing ones or self-written) is not specified. (2) There lacks a commonly used stock dataset (like MNIST and ImageNet in computer vision), while self-built ones are seldom released. (3) The experiment settings are always not presented with sufficient details, as described in the Introduction section.

3 RL Modeling of Stock Trading

In this section, we describe how we model the multi-stock trading process and analyze the feasibility of RL-based methods. Due to space limitation, we will skip the introduction of RL's basic theory and the mechanism of A2C and PPO algorithms.

3.1 Problem Description

This article is mainly focused on how to use the deep reinforcement learning method to implement automated multi-stock trading and evaluate its performance. That is, creating an agent that can dynamically distribute a certain amount of money among a set of stocks over a period of time in order to maximize return. The target of trading is a fixed group of stocks, so stock selection is not considered in training. The number of target stocks is denoted by N in the sections that follow.

According to RL theory, the relationship between investors and the stock market can be viewed as an interaction between an agent and the environment. The state of the environment consists of two parts: the stock's state and how the investor is participating in the market. The former can be depicted with stock price, trading volume/amount, and indicators. We use close price as daily price. The selection between volume or amount is explained in Action Space part. We pick three most commonly used technical indicators, namely Moving Average Convergence/Divergence (MACD), Stochastic oscillator (KD), and Relative Strength Index (RSI), plus the stock price as the state representation. Since the indicators already include information about historical prices implicitly, we do not add historical stock prices as part of the state like many similar works do. The latter can be directly represented by the investor's holding of each stock. On every trading day, the investor has three options of action for each stock: buy, hold, and sell. The reward for an investor is the change of its total assets, i.e., balance plus the value of stock held, over one trade date.

3.2 Mathematical Presentation

State Space. The state of day t is represented as a $6N+1$-dim vector: $s(t) = [b_t, h_t, p_t, M_t, K_t, D_t, R_t]$. Each component is explained as follows:

- $b_t \in \mathbb{R}_{\geq 0}$: agent's balance, non-negative scalar.
- $h_t \in \mathbb{R}^N_{\geq 0}$: agent's holding of each stock, N-dim non-negative vector.
- $p_t \in \mathbb{R}^N_{\geq 0}$: each stock's price, N-dim non-negative vector.
- $M_t \in \mathbb{R}^N$: each stock's DIF value in MACD indicator, N-dim vector.
- $K_t \in [0, 100]^N$: each stock's K value in KD indicator, N-dim vector in range $[0, 100]$.
- $D_t \in [0, 100]^N$: each stock's D value in KD indicator, N-dim vector in range $[0, 100]$.
- $R_t \in [0, 100]^N$: each stock's RSI indicator, N-dim vector in range $[0, 100]$.

Action Space. We define the action space as $[-1, 1]^N$ because most RL algorithms require or learn best with a Gaussian distribution. Each value in one dimension represents the buy/hold/sell action for one stock, where positive value means buying and negative value means selling. The vector will first be multiplied by a constant factor *amp* when sent into the environment and then be converted to the actual action. Considering that the magnitude of each stock's price may vary widely, we explain the action value as the amount of money to be traded, instead of volume to be traded. The amount will then be divided by the stock price and rounded down to get the volume. If the value is negative (selling), it will be clipped to the agent's holding amount as the agent can't sell more than that. By default, *amp* factor is set to $\frac{b_0}{N}$, where b_0 is the initial fund.

Reward Function. We define the reward signal as the daily return, in other words, the change of the total asset over one trade day. The total asset consists of the balance and stock hold's value, so the function can be written as:

$$r(s_t, a_t, s_{t+1}) = (b_{t+1} + h_{t+1} \cdot p_{t+1}) - (b_t + h_t \cdot p_t) \tag{1}$$

3.3 Trading Details

The mechanism of the real world stock market is rather complex. We make following assumptions to moderately simplify the case which is similar to previous work [16]:

1. Amount of fund. Assume the amount of the agent's trading is small enough so that it won't affect the stock's price. This is the prerequisite that an agent can learn and be tested on historical stock data.
2. Market liquidity. Assume all of the agent's trading can be done immediately at the day's close price.
3. Transaction cost. The Chinese stock market specifies a series of transaction fee items. We only consider the stamp duty, which is at the highest rate: 0.1%, paid by the seller.
4. No debt. Assume the agent's balance must be non-negative at any time. This is intended to prevent agent from learning unorthodox strategies. The environment will process the agent's actions in dimensional order, and if one action tries to buy an amount greater than the agent's balance, it will be clipped.

3.4 Feasibility Analysis of RL-Based Stock Trading

The RL approach has several natural advantages for stock trading:

1. Low data cost. A large amount of historical stock data is available on many financial sites. The data is both cheap and quality-guaranteed.
2. Adequate reward signal. Sparse reward signals are a great challenge in RL as they make it difficult to learn the difference in value between states. For example, when dealing with some games, it is definitely not enough to only give the winning/losing step a $+1/-1$ reward. However, for stock trading problems, the reward signal can naturally be defined as the daily return, which has a precise value on every step, and the value is likely not to be zero in most situations.
3. Aligned reward and goal. In RL, the optimization objective is defined as a reward function, but in many cases, this needs some conversion from the initial goal. For example, if the goal is to make a robot pick up a cube and put it on another one, the reward function might be defined as the distance between the cube's bottom surface and the ground. However, this may sometimes lead to unpleasant learning results, e.g., the robot might learn to just turn the cube bottom-side up. This is not an issue for stock trading, as the reward

we defined is exactly the initial goal: maximizing the investment return. DL-based methods for stock trading also suffer from this problem, as described in the introduction.

The shortcoming for RL and also other ML-based approaches for stock trading is that there are too many factors affecting the stock price, while the model can only use a very limited set as input to learn. We hope for more work on methods of extracting highly featured factors, as we believe such work has the potential to significantly improve the model's performance.

4 Experiments

4.1 Stock Dataset

Stock Selection. We select two sets of data from the Chinese stock market. The first one is the 17 constituent stocks of the SSE (Shanghai Stock Exchange) 50 index, which represents Chinese large-capitalization stocks. The training period ranges from 2010/01/01 to 2017/12/31, and the testing period ranges from 2018/01/01 to 2021/12/31. The second one consists of 10 stocks from GEI (Growth Enterprise Index), which represents small-capitalization yet growing stocks, whose prices are more volatile. Because some stocks in this group are listed in 2011, we set the training period as 2012/01/01 to 2017/12/31, and the test period is the same.

Data Processing. We acquire all the stock data from Tushare. Due to the relatively long training and testing period (12 years in total), many of the stocks have missing data on some trade dates because of temporary suspension or other reasons. Therefore, we drop all the trade dates where at least one stock doesn't have data. This processing method loses about 10% of the data, which is still acceptable.

4.2 Methodology

As investigated in [8], RL model's performance may vary significantly between repeated training only because the random seed has shifted, especially when the environment that the agent interacts with is unstable. Therefore, we deem that it is of great importance that the model should be trained and tested multiple times to draw a credible conclusion about its real performance. In this article, all the experiments are repeated 10 times, and the average performance along with standard deviation is reported. Moreover, to verify that the trading strategy the agent learned is reasonable, we plot the total asset curve and asset distribution race chart to observe how the agent conducts trading action to earn the return. To the best of our knowledge, this is the first work to do such analysis regarding RL for stock trading.

We use stable-baselines3 as the algorithm implementation. This article is not aimed at exploring how to find the best hyperparameters for the RL model, so we just use the default ones provided. The only parameter we change is the most basic for RL: the number of training time-steps, or how many steps an agent can interact with the environment to learn. Usually this is set to a value between 10^4 and 10^7 in RL training, so we choose eight values: 10K, 25K, 50K, 100K, 250K, 500K, 1M, 2M (K = thousand, M = million) to test the model. This parameter is barely reported in previous works, which makes reproducing the results very difficult if the source code is not released.

We don't set a validation period in our experiments. Traditional ML problems are usually considered to be independently identically distributed, so that the model's actual performance can be observed and tuned on a separate validation set. For stock trading, however, the environment (market) shows regularity in the long term but is rather chaotic in the short term. Tuning hyperparameters or selecting the best-performing model on a relatively short validation period may not be effective for improving test period performance.

One advantage of RL is that it can keep learning to adapt to the environment, while the environment might also be changing by itself. For tasks like playing video games, since the game rules are given first, the environment's characteristics are fixed, so there is actually no difference between training and testing. The stock market environment, however, is volatile and the trend of the training and testing period might differ. Considering this, we think it is beneficial to adopt an online learning-like method, i.e., to stop and continue to train the model using new-coming stock data whenever a time interval has passed on the test period, which we call a track training strategy. The process is shown in Fig. 1. In the experiment, we apply this strategy to the best-performed A2C and PPO models to observe the effect. We set the retraining interval as one quarter. Since retraining only uses last quarter's data, which is much shorter compared to the whole training period, we test the model with three retraining steps: 5K, 10K, and 25K.

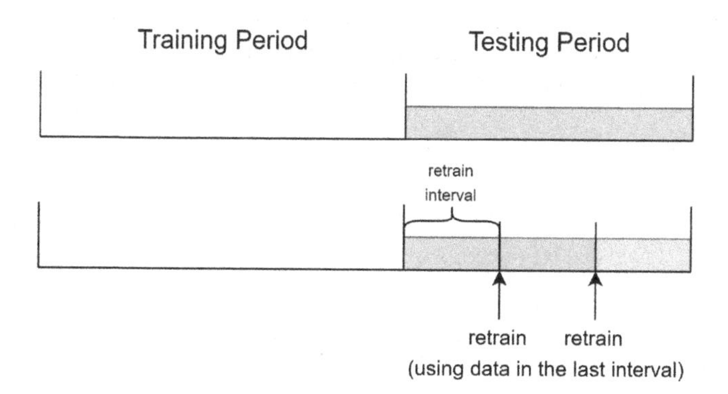

Fig. 1. The comparison of training-testing and track training. Different color indicates the model making actual trading action is different.

4.3 Results

Baseline. We use a simple average holding strategy as the baseline: we divide the initial fund equally into N parts to buy each stock on the first trade day, and do nothing later. The result of this strategy is deterministic: the annual return rate is 12.70% and 42.27% on two datasets.

Model Performance. Table 1 shows the performance of A2C and PPO agents on the SSE17 dataset. Both A2C and PPO agents significantly out-perform the baseline strategy when trained with $\geq 10^5$ time-steps. However, we notice that the average performance doesn't increase monotonically with time-steps like usual RL problems.

Table 1. A2C and PPO agent's performance on SSE17 dataset.

Train steps	Baseline	A2C agent	PPO agent
10K	12.70(±0)%	10.74(±7.56)%	7.26(±1.49)%
25K		8.43(±6.41)%	8.32(±2.07)%
50K		15.12(±10.45)%	6.92(±1.17)%
100K		**22.41(±11.92)%**	8.32(±3.03)%
250K		13.89(±5.79)%	19.03(±7.31)%
500K		12.62(±6.62)%	21.77(±11.99)%
1M		13.11(±2.55)%	**31.68(±11.35)%**
2M		16.33(±3.25)%	25.45(±16.11)%

For A2C, the performance goes up first, then drops at 250K, and later goes up again, while the standard deviation keeps going down. To prove that the agent is taking reasonable action, we plot the asset curve and asset distribution chart of the best-performed 2M-steps model, shown in Fig. 2 and Fig. 3.

Fig. 2. The asset curve of the best-performed 2M-steps A2C model.

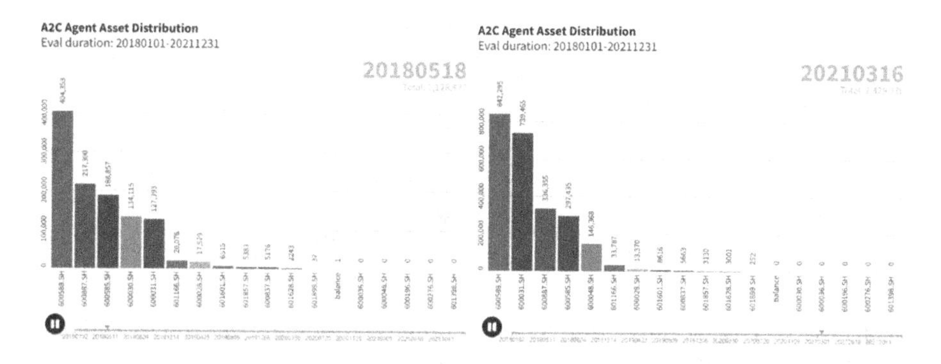

Fig. 3. The asset distribution status of the best-performed 2M-steps A2C model on two trade dates in the test period.

It can be seen that the A2C agent behaves well, especially in a bullish market trend, and doesn't lose too much in a fluctuating or bearish trend. Figure 3 shows that the agent keeps investing in multiple stocks, and has changed the proportion of holdings over the test period. These evidences prove that the A2C agent has learned a profitable and reasonable trading strategy.

For PPO, the performance peaks at 1M but then goes down at 2M-steps, while the standard deviation keeps increasing. To explain the result, we also plot the asset distribution chart and find that the PPO model tends to focus too much on one stock. That causes the large deviation, as the return rate highly depends on one stock's trend. We find decreasing the *amp* factor, i.e., limiting the maximum trading amount on each date when testing, can alleviate this issue, although it can cause the return rate to drop by about 6% (but still 9% higher than baseline). Figure 4 shows the different asset distribution of the same model on the same date with different *amp* value (the default $\frac{b_0}{N} \approx 58823$ and 5000). Nevertheless, if the agent is both trained and tested with $amp = 5000$, though the standard deviation narrows significantly, the return rate also drops below the baseline, even at 2M-steps (the result is 11.97($\pm$4.52)%). This suggests that the PPO agent's learning is sensitive to this parameter, and it is better to use different value for training and testing.

We then employ the track training strategy on the best-performed 2M-steps A2C and PPO models. The result is shown in Table 2.

The strategy enhances the A2C and PPO model's performance by 6.5% and 2.5%, with only about 15% extra training time. This is far less expensive than performing a hyperparameter grid search or continue to increase the train step in a million level. It is also worth noting that the performance of track training is highly related to the retraining step. A large value can cause the model to perform even worse than the original. This is probably related to the catastrophic forgetting issue.

The above experiments are repeated on the second GEI10 dataset. Table 3 shows the A2C and PPO agents' performance. As the stocks that form this dataset are more volatile, we can see that both agents' standard deviation under

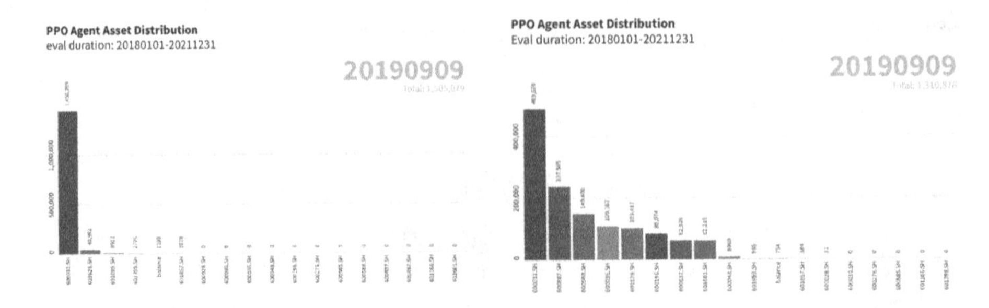

Fig. 4. Best-performed 2M-steps PPO model's different asset distribution with $amp \approx 58823$ and $amp = 5000$.

Table 2. Best-performed 2M-steps A2C and PPO model's performance with track training on SSE17 dataset.

Retrain steps	Baseline	A2C model	PPO model
0 (Original model)	12.70(±0)%	25.32%	21.65%
5K		29.59%	22.71%
10K		**31.78%**	23.98%
25K		25.19%	**24.24%**

all parameters are larger. On the other hand, the average performance of both agents are still significantly above the baseline. The result suggests that it is necessary to use multiple trained models to make decisions together to obtain a good average return. The outcome of a single training is not stable enough.

Table 3. A2C and PPO agent's performance on GEI10 dataset.

Train steps	Baseline	A2C model	PPO model
10K	42.27(±0)%	34.93(±16.31)%	43.47(±17.71)%
25K		33.15(±23.31)%	41.94(±26.09)%
50K		**50.23(±31.25)%**	38.38(±14.38)%
100K		46.26(±19.81)%	38.08(±15.74)%
250K		30.91(±14.55)%	**68.88(±21.78)%**
500K		42.53(±14.55)%	56.18(±19.86)%
1M		32.72(±23.93)%	50.55(±24.97)%
2M		38.62(±19.72)%	47.72(±23.71)%

The result of adding track training strategy to the best 2M-steps model is shown in Table 4. The strategy is still effective on this dataset and enhances both models' performance by about 10% at 5K retrain steps.

Table 4. Best-performed 2M-steps A2C and PPO model's performance with track training on GEI10 dataset.

Retrain steps	Baseline	A2C model	PPO model
0 (Original Model)	42.27(±0)%	73.85%	97.20%
5K		**84.77%**	**108.21%**
10K		65.69%	84.45%
25K		53.87%	78.82%

5 Conclusion and Future Works

In this article, we propose an RL-based modeling method of multi-stock trading and demonstrate that both the A2C and PPO algorithms can earn an investment return significantly higher than the baseline. However, the good performance is based on the average result of multiple training, whereas the outcome of a single training is not stable enough. We also show that employing a track training strategy can further enhance both models' performance with little extra cost, but the retraining step needs to be chosen carefully, otherwise the result may be worse than the original.

We hope that this work will serve as a foundational study of an RL-based approach to quantitative stock trading. In the future, we're expecting to see more works in two major directions: (1) The algorithm and model. A2C and PPO are popular yet relatively old algorithms. More advanced ones like TD3, SAC can be employed and may perform better, though they might require more subtle tuning. The design of network structures can also be explored. (2) State modeling. In this article, we use three technical indicators and the stock price to represent the market's state. If indicators from fundamental analysis and investor sentimental analysis could be integrated, the model is likely to perform better as it can see more featured information. This is still an open and active research area in ML for finance. We are also glad to discover that there are emerging works, such as FinRL library [10], that focus on developing a general-purpose open-source framework for employing deep RL in finance. These works are essential for attracting more researchers into this area and helping to create more shared datasets and comparable experiment results, which is the key to pushing current research to a higher level.

References

1. Azhikodan, A.R., Bhat, A.G.K., Jadhav, M.V.: Stock trading bot using deep reinforcement learning. In: Saini, H.S., Sayal, R., Govardhan, A., Buyya, R. (eds.) Innovations in Computer Science and Engineering. LNNS, vol. 32, pp. 41–49. Springer, Singapore (2019). https://doi.org/10.1007/978-981-10-8201-6_5
2. Chong, E., Han, C., Park, F.C.: Deep learning networks for stock market analysis and prediction: methodology, data representations, and case studies. Expert Syst. Appl. **83**, 187–205 (2017)

3. Dang, Q.-V.: Reinforcement learning in stock trading. In: Le Thi, H.A., Le, H.M., Pham Dinh, T., Nguyen, N.T. (eds.) ICCSAMA 2019. AISC, vol. 1121, pp. 311–322. Springer, Cham (2020). https://doi.org/10.1007/978-3-030-38364-0_28
4. DaoYi, X.: Chinese stock market capitalization to GDP over the years. http://www.xindaoyi.com/market-value-of-gdp/. Accessed 01 June 2022
5. Fischer, T.G.: Reinforcement learning in financial markets-a survey. Technical report, FAU Discussion Papers in Economics (2018)
6. François-Lavet, V., Henderson, P., Islam, R., Bellemare, M.G., Pineau, J.: An introduction to deep reinforcement learning. arXiv preprint arXiv:1811.12560 (2018)
7. Gudelek, M.U., Boluk, S.A., Ozbayoglu, A.M.: A deep learning based stock trading model with 2-D CNN trend detection. In: 2017 IEEE Symposium Series on Computational Intelligence (SSCI), pp. 1–8. IEEE (2017)
8. Henderson, P., Islam, R., Bachman, P., Pineau, J., Precup, D., Meger, D.: Deep reinforcement learning that matters. In: Proceedings of the AAAI Conference on Artificial Intelligence, vol. 32, no. 1 (2018). https://doi.org/10.1609/aaai.v32i1.11694, https://ojs.aaai.org/index.php/AAAI/article/view/11694
9. Li, Y., Ni, P., Chang, V.: Application of deep reinforcement learning in stock trading strategies and stock forecasting. Computing **102**(6), 1305–1322 (2020)
10. Liu, X.Y., Yang, H., Chen, Q., Zhang, R., Yang, L., Xiao, B., Wang, C.D.: FinRL: a deep reinforcement learning library for automated stock trading in quantitative finance. arXiv preprint arXiv:2011.09607 (2020)
11. Meng, T.L., Khushi, M.: Reinforcement learning in financial markets. Data **4**(3), 110 (2019)
12. Pricope, T.V.: Deep reinforcement learning in quantitative algorithmic trading: a review. arXiv preprint arXiv:2106.00123 (2021)
13. Vezeris, D., Karkanis, I., Kyrgos, T.: AdTurtle: an advanced turtle trading system. J. Risk Finan. Manag. **12**(2), 96 (2019)
14. Wikipedia contributors: Stock — Wikipedia, the free encyclopedia (2022). https://en.wikipedia.org/w/index.php?title=Stock&oldid=1102819292. Accessed 01 June 2022
15. Wu, X., Chen, H., Wang, J., Troiano, L., Loia, V., Fujita, H.: Adaptive stock trading strategies with deep reinforcement learning methods. Inf. Sci. **538**, 142–158 (2020)
16. Yang, H., Liu, X.Y., Zhong, S., Walid, A.: Deep reinforcement learning for automated stock trading: an ensemble strategy. In: Proceedings of the First ACM International Conference on AI in Finance, pp. 1–8 (2020)
17. YCharts: US total market capitalization as. https://ycharts.com/indicators/us_total_market_capitalization. Accessed 01 June 2022

InDNI: An Infection Time Independent Method for Diffusion Network Inference

Guoxin Chen, Yongqing Wang(✉), Jiangli Shao, Boshen Shi, Huawei Shen, and Xueqi Cheng

Data Intelligence System Research Center, Institute of Computing Technology, Chinese Academy of Sciences, Beijing, China
{wangyongqing,shaojiangli19z,shiboshen19s,shenhuawei,cxq}@ict.ac.cn

Abstract. Diffusion network inference aims to reveal the message propagation process among users and has attracted many research interests due to the fundamental role it plays in some real applications, such as rumor-spread forecasting and epidemic controlling. Most existing methods tackle the task with exact node infection time. However, collecting infection time information is time-consuming and labor-intensive, especially when information flows are huge and complex. To combat the problem, we propose a new diffusion network inference algorithm that only relies on infection states. The proposed method first encodes several observation states into a node infection matrix and then obtains the node embedding via the variational autoencoder (VAE). Nodes with the least Wasserstein distance of embeddings are predicted for existing propagation edges. Meanwhile, to reduce the complexity, a novel clustering-based filtering strategy is designed for selecting latent propagation edges. Extensive experiments show that the proposed model outperforms the state-of-the-art infection time independent models while demonstrating comparable performance over infection time based models.

Keywords: Diffusion network inference · Variational autoencoder · Wasserstein distance

1 Introduction

The topology of a diffusion network reveals how information is propagated among users, intuitively illustrating potential information propagation paths. Promoting and preventing future diffusion on the network are of great importance [23]. Generally, observing the influence relationship between users in the real scenarios is difficult, thus researchers try to recover the relations with the historical propagation [9], which is known as the diffusion network inference task.

According to whether the node infection time is used or not, most of the existing diffusion network inference methods can be divided into two categories: the infection time dependent and infection time independent algorithms [11]. The infection time dependent algorithm mostly assume that previously infected nodes are potential parents of subsequently infected nodes. It infers the influence

Y. Chang and X. Zhu (Eds.): CCIR 2022, LNCS 13819, pp. 63–75, 2023.
https://doi.org/10.1007/978-3-031-24755-2_6

relationship between nodes by constructing and maximizing different likelihood functions based on the exact infection time [15]. However, in some realistic scenarios, monitoring and recording the node infection time is labor demanding and time consuming, limiting the performance and application of the algorithm. Therefore, works on reconstructing the diffusion network structure without infection time have emerged. Two existing methods [2,8] attempt to learn influence relationships between nodes from all fixed-length path trajectories or from initial and resulting infected node sets. However, the former requires obtaining all fixed-length path trajectories, the latter requires prior knowledge of the number of edges in the diffusion network. These are difficult to obtain in practice.

In order to address the above problems, we proposes a novel infection time independent network inference method called InDNI (An Infection Time **In**dependent Method for **D**iffusion **N**etwork **I**nference). InDNI takes final states of nodes (infected or not) in different propagation processes as inputs, which is easier to obtain than the exact infection time or infection sequences. During information propagation, user pairs with follower relationships tend to exhibit similar behaviors (infected or not infected at the same time). Therefore, the final infection state of a node can well reflect the influence relationship between nodes. InDNI utilizes variational auto-encoders [4] to extract the behavioral characteristics of nodes. Meanwhile, wasserstein distance [24] is utilized to measure the similarity between node pairs and further infer the adjacent edge relationship between nodes. Furthermore, to address the problem that nodes rarely involved in propagation will be aggregated within the embedding space and be misjudged as having adjacent edges, InDNI excludes node pairs with very low probability of adjacent edges by examining the correlation of the infection mutual information (IMI) metric [9] between nodes. In summary, the main contributions of this paper are summarized as follows:

- We propose a novel diffusion network inference algorithm named InDNI. Compared to existing algorithms, InDNI relies only on the final infection status of nodes, which is easier to obtain in practice.
- An effective filtering strategy based on infection mutual information is proposed for not only accelerating the prediction process but also reducing the bias led by infrequent nodes.
- Experimental results on synthetic and real-world dataset demonstrates that InDNI outperforms the state-of-the-art infection time independent models while showing comparable performance over infection time based models.

2 Related Work

According to whether the infection time is utilized, diffusion network inference can be divided into the following two categories: (1) The infection time-based algorithm; (2) The infection time-independent algorithms.

The Infection Time-Based Algorithm. Most of the infection time-based network inference algorithms are based on the following assumptions: nodes that are sequentially infected within a time interval have influence relationships, and previously infected nodes are considered as potential parents of subsequently infected nodes. Therefore, these methods rely on explicit temporal information for each node. Representative algorithms are: InfoPath [21], NetINF [7], REFINE [13], etc. Depending on the differences in the principles, this types of methods can be subdivided into: the convex programming-based approaches, the submodularity-based approaches, and the embedding-based approaches. Although the infection time-based algorithm is effective in some scenarios to solve the network inference problem, it is a very difficult task to obtain the data with the accurate infection time during the propagation. This undoubtedly challenges the applicability of the algorithm.

The Infection Time-Independent Algorithms. To solve the problem of time collection difficulties, some works try to learn the influence relationships between nodes from diffusion path traces (the Path approach [8]), based on lifting effects (the LIFT approach [2]) or the node infection sequence (the DeepINFER approach [12]). The PATH approach requires obtaining all fixed-length path trajectories. The LIFT approach requires a priori knowledge of the number of edges in the diffusion network, otherwise it infers a fully connected graph. All these methods have strong a priori assumptions that are not conducive to the application of the algorithm. Subsequently, DeepINFER proposes to compare the infection sequence context of a node with the textual context and learn the node representation by Skip-Gram model [20]. It gets rid of the dependence on infection time to a certain extent. However, due to the hysteresis of information propagation, there is not necessarily a direct influence relationship between the context node and the central node, which will make the model mistakenly believe that there is a high probability of adjacent edges between the context node and the central node.

3 Problem Statement

The diffusion network can be represented as a graph $G = \{V, E\}$, where $V = \{v_1, v_2, ..., v_n\}$ refers to the set of n nodes, and E denotes the set of m edges between nodes. In the diffusion network inference problem, the set of nodes is given and the set of edges is unknown and needs to be inferred. In this paper, we assume that the propagation result only contains the final infection state of the node, not the infection time or sequence. Therefore, the formal definition of our problem statement is given (Table 1 gives the main symbols in this paper):

Given: *On the diffusion network G, given a set of node infection status results $C = \{C^1, C^2, \ldots, C^k\}$ during k times of historical propagation. The element $C^l = \{c_1^l, c_2^l, \ldots, c_n^l\}$ is an n-dimensional vector (n is the number of nodes in G) which records the final infection status of each node in the lth propagation, and $c_i^l = \{0, 1\}$ (0 refers to not infected, 1 refers to infected).*

Infer: *The unknown edge set E of the diffusion network G.*

Table 1. A brief summary of notions.

Symbol	Description
G	A diffusion network
V, E	The node set and edge set of network G
n, m	The number of nodes and edges of network G
C	The observed infection status of nodes in G during historical propagation
X	Node infection matrix extracted from C
S_{pair}	The set of candidate node pairs, i.e., node pairs that may have adjacent edges

4 InDNI Algorithm

From the perspective of graph representation learning, the basic idea of the diffusion network inference is to learn the node representation according to the result of propagation, and judge whether there is an edge based on the similarity between nodes. Therefore, InDNI extracts the behavioral features of nodes in the process of reconstructing the infection state of nodes with the help of VAE. Then, InDNI describes the similarity between nodes through the Wasserstein distance for diffusion network inference. Meanwhile, in order to solve the problem of aggregation of nodes that are isolated or rarely involved in propagation in the embedding space, InDNI introduces IMI metrics for preliminary filtering to obtain candidate pairs of nodes that may have adjacent edges.

4.1 Node Representation Learning

Node Initial Features. Based on the assumption that adjacent nodes often have similar behaviors in information propagation, the following definition of the node infection matrix is given as input to the model:

Definition 1 (Node Infection Matrix). *On a diffusion network G containing n nodes, the set of node infection states $C = \{C^1, C^2, \ldots, C^k\}$ during k historical propagation is given. Extracting the node infection matrix from C is defined as $X \in \mathbb{R}^{n \times k}$. The i-th row and j-th column of the matrix X represent the infection of node v_i in the j-th propagation:*

$$X_{ij} = \begin{cases} 1, \ if \ c_i^j = 1 \\ 0, \ if \ c_i^j = 0 \end{cases} \tag{1}$$

The node infection matrix X only depends on the infection results of the propagation, which represents the behavioral tendency of nodes in this group of propagation.

Variational Autoencoder. InDNI maps node infection states to a low-dimensional space through a Variational Autoencoder (VAE) containing multiple non-linear layers, aiming to extract dense behavioral feature vectors of nodes from the node infection matrix X. VAE utilizes Gaussian distribution to describe the probability distribution of node features, whose powerful information extraction capability has been demonstrated in several fields such as graph representation learning [24] and graph generation [14].

Loss Function. The loss function of VAE mainly consists of two parts: reconstruction loss and distribution loss [4], as shown by the Eqs. 2. In the formula, $p(Z) = \prod_i p(z_i) = \prod_i \mathcal{N}(z_i|0, I)$ is a Gaussian prior distribution of the latent variable Z, and $Q(Z|X, \Phi)$ is the Encoder, Φ is the parameter of the Encoder, $P(\hat{X}|Z, \Theta)$ is the Decoder, Θ is the parameter of the Decoder, and KL refers to the KL divergence.

$$\mathcal{L} = E_{Q(Z|X,\Phi)}\left[\log P(\hat{X}|Z, \Theta)\right] - \mathrm{KL}\left[Q(Z|X, \Phi)||p(Z)\right] \tag{2}$$

The reconstruction loss hopes that the features extracted by the Encoder can restore the input X well. It is worth mentioning that the node infection matrix X is usually sparse. In order to avoid the model from over-learning the sparse part, inspired by existing work, in the reconstruction loss, we assign more penalties to the loss caused by non-zero elements than the loss caused by zero elements. So the reconstruction loss can be defined as:

$$E_{Q(Z|X,\Phi)}\left[\log P(\hat{X}|Z, \Theta)\right] = \sum_{i=0}^{n} \|(X_i - \hat{X}_i) \odot P_i\|_2^2 = \|(X - \hat{X}) \odot P\|_F^2 \tag{3}$$

where $\odot$ denotes the Hadamard product, and the elements of row i and column j of the weight matrix P are defined as follows:

$$P_{ij} = \begin{cases} \rho, \ if\ X_i^j = 1 \\ 1, \ if\ X_i^j = 0 \end{cases} \tag{4}$$

where ρ is the hyperparameter of the model, which is used to adjust the weight of the reconstruction loss of non-zero elements. In the experimental part, the effect of different ρ on the results is explored.

The distribution loss can be thought of as a regularizer. Its goal is to reconstruct a meaningful output X even when the latent variable Z is sampled from a priori distribution $p(Z)$. We assume that the latent variable Z follows a Gaussian distribution, as shown by the Eqs. 5.

$$\begin{aligned} \mathrm{KL}\left[Q(Z|X, \Phi)||p(Z)\right] &= \mathrm{KL}\left[\mathcal{N}(\mu(X), \Sigma(X))||\mathcal{N}(0, I)\right] \\ &= \frac{1}{2}(\mathrm{Tr}(\Sigma(X)) + \mu(X)^T\mu(X) - k + \log(\det(\Sigma(X)))) \end{aligned} \tag{5}$$

where k denotes the dimensionality of the distribution features, Tr is the trace of the matrix, det is the determinant of the matrix, and the parameters μ and Σ of the Gaussian distribution can be obtained by fitting a neural network while Σ is constrained to be a diagonal matrix.

4.2 Similarity Measure

After obtaining the node representation, a suitable distance metric is needed to describe the similarity between different nodes. Extensive work [24] has demonstrated that the Wasserstein distance is a good measure of the distance between two distributions. It does not suffer from problems similar to the KL divergence or JS divergence that give meaningless or constant results for two distributions that do not overlap at all. This ensures that it can stably describe the similarity between nodes. The qth Wasserstein distance between two probability distributions P_1 and P_2 is defined as:

$$W_q(P_1, P_2) = (\inf \mathbb{E}\left[\mathrm{d}(X, Y)^q\right])^{1/q} \tag{6}$$

where $\mathbb{E}[Z]$ denotes the expected value of a random variable Z and the infimum is taken over all joint distributions of the random variables X and Y with marginals P_1 and P_2 respectively.

However, the general form of Wasserstein distance is limited by the large computational cost and it is difficult to apply it directly to the problem in this paper. To reduce the computational cost, in this paper, the 2nd Wasserstein distance (abbreviated as W_2) has a closed form solution to speed up the computational process since we use a Gaussian distribution as the node representation. In the meanwhile, we focus on diagonal covariance matrices [6,22], i.e., $\Sigma_1\Sigma_2 = \Sigma_2\Sigma_1$. Thus, Eq. 6 can be simplified as Eq. 7 [24]. The computational complexity of W_2 scales linearly with the embedding dimension.

$$W_2(\mathcal{N}(\mu_1, \Sigma_1), \mathcal{N}(\mu_2, \Sigma_2)) = (\|\mu_1 - \mu_2\|_2^2 + \|\Sigma_1^{1/2} - \Sigma_2^{1/2}\|_F^2)^{1/2} \tag{7}$$

4.3 Filtering Candidate Node Pairs

There are often isolated nodes or nodes that rarely participate in the propagation in the diffusion network. Since their infection status is basically all zeros, these nodes cluster within the embedding space when projected onto the embedding space via the VAE. Obviously, by the above approach alone, the algorithm would naturally assume that there are adjacent edges between these nodes, which do not actually exist. To solve the above problem, InDNI introduces the Infection-MI metric for initial filtering to obtain the set of candidate node pairs for which adjacent edges may exist.

Infection-MI. Based on the assumption that neighboring nodes tend to have similar behaviors during information propagation, we argue that the higher the behavioral correlation, the higher the probability of the existence of adjacent edges. Inspired by the existing work [9], we introduce the Infection-MI metric (abbreviated as IMI) based on mutual information to measure node behavioral correlation.

$$\mathrm{IMI}(v_i, v_j) = \mathrm{MI}(v_i = 1, v_j = 1) - \mathrm{MI}(v_i = 1, v_j = 0) - \mathrm{MI}(v_i = 0, v_j = 1) \quad (8)$$

According to the above definition, when the behaviors of nodes v_i and v_j are highly negatively correlated, i.e., the values of $\mathrm{MI}(v_i = 1, v_j = 0)$ or $\mathrm{MI}(v_i = 0, v_j = 1)$ is significantly large, $\mathrm{IMI}(v_i, v_j)$ tends to be negative. When the behavior of v_i and v_j tend to be independent, the values of $\mathrm{MI}(v_i = 1, v_j = 1)$, $\mathrm{MI}(v_i = 1, v_j = 0)$ and $\mathrm{MI}(v_i = 0, v_j = 1)$ are very small and the value of IMI is close to 0. When the behavior of A and B have a high correlation, IMI is a relatively large positive value. Therefore, IMI can well reflect the correlation of behaviors between nodes.

Kmeans-Based Filtering Method. In diffusion networks, each node v_i usually contains only a limited number of parents that tend to have a large positive correlation with node v_i. Except for a few nodes that have negative correlation with v_i, most of the nodes in the network have no influence relationship on v_i, which leads to a compact cluster with a mean value close to 0 for the IMI metric. To avoid setting hyperparameters and to distinguish weak positive correlation from positive correlation, we introduce a filtering method based on Kmeans clustering [10] to filter out the set of node pairs S_{pair} with possible adjacent edges based on IMI, with the following procedure.

Step 1: Calculate the IMI value between each node in the network and remove the node pairs in which the IMI value is negative.

Step 2: Executing the Kmeans algorithm for the remaining pairs of nodes (one-dimensional clustering), where $K = 2$, fixing one of the cluster centers to 0 and initializing the other cluster center to the maximum of all IMI values, iterating continuously until stability.

Step 3: The node pairs contained in clusters with non-zero cluster centers are treated as the set of candidate node pairs, denoted as S_{pair}, i.e., the node pairs with possible adjacent edges.

4.4 Network Inference

Once the representation of each node and the set of candidate node pairs S_{pair} are obtained, the next step is to infer the set of edges E of the diffusion network [12]. We believe that the distributional difference between node pairs with adjacent edges should be smaller, and the distributional difference between node pairs without adjacent edges should be larger. Therefore, we can infer whether there is an edge by computing the similarity of candidate node pairs, which is expressed formally as Eq. 9.

$$E = \{(u, v) : W_2(u, v) \leq \tau, (u, v) \in S_{pair}\} \quad (9)$$

Up to this point, all the steps have been clarified and the overall flow of InDNI is summarized in Algorithm 1.

Algorithm 1. The InDNI Algorithm

Input: The node set V, the node infection matrix X.
Output: The edge set E.
1: **for** $i = 1$ to n **do**
2: **for** $j = i + 1$ to n **do**
3: Calculate IMI(v_i, v_j) by Eq. 8.
4: **end for**
5: **end for**
6: Based on IMI, S_{pair} is obtained by Kmeans filtering method.
7: The node infection matrix X is used as input and the node representation is obtained by iteratively optimizing the VAE parameters according to Eq. 2.
8: **for** $(v_i, v_j) \in S_{pair}$ **do**
9: Calculate $W_2(v_i, v_j)$ by Eq. 7.
10: **end for**
11: By Eq. 9, infer the edge set E.

5 Experiments

5.1 Experimental Setup

Datasets. To comprehensively observe the capabilities of the InDNI algorithm, we conduct experiments on different scenarios on synthetic and real datasets. The basic statistics of each dataset are shown in Table 2. We first consider the Kronecker model [16] to simulate a real-world diffusion network, which can generate different network structure that approximates the real-world scenario based on different parameter inputs. In this paper, we consider the three most common network structures: Random [5] (parameter matrix: [0.5, 0.5; 0.5,0.5]), Hierarchical [3] ([0.9, 0.1; 0.1, 0.9]) and Core-periphery [17] ([0.9, 0.5; 0.5, 0.3]).

Table 2. A brief description about datasets

Type	Networks	n	m	k
Synthetic datasets	Kronecker Random	512	1024	5000
	Kronecker Hierarchical	512	1024	5000
	Kronecker Core-Periphery	512	1024	5000
Real world datasets	Dolphin	62	318	1000
	Polblogs	1490	33430	10000
	Facebook	4039	88234	50000

Although the synthetic model can generate the desired network structure by controlling different parameters, which facilitates the in-depth analysis of the algorithm performance. However, synthetic datasets are not really a substitute for real-world propagation processes, which are often influenced by a large number of external factors and cannot be really accurately described. Therefore, to explore the capability of the algorithm in real-world scenarios, we conducted extensive experiments on real datasets as follows: (1) **Dolphin:** [19] A small undirected social network of frequent contact between 62 dolphins in a community near the New Zealand Strait; (2) **Polblogs:** [1] A network of hyperlinks between blogs about American politics recorded in 2005; (3) **Facebook:** [18] A social network from Facebook that has anonymized user ids.

Baseline Methods. Among the existing network inference algorithms, it is mainly divided into infection time-based and infection time-independent network inference algorithms. The accuracy of infection time-based inference algorithms tends to be higher than that of infection time-independent inference algorithms, because the former uses more information and requires higher quality of data. We selected the most representative algorithms of the two major classes as the benchmark algorithms for our experiments, listed as InfoPath [21], Netinf [7], REFINE [13], DeepINFER [12]. To evaluate the accuracy of InDNI in diffusion network inference, we report the F1-score [9] as an evaluation metric.

5.2 Results and Discussion

Diffusion Network Inference. Following the above settings, we compare the performance of the algorithms on six datasets, the results are presented in Table 3. As seen from the table, (1) the accuracy of infection time-based inference algorithms tends to be higher than that of infection time-independent inference algorithms, because the former uses more information and requires higher quality of data. (2) InDNI outperforms DeepINFER on all datasets, while approaching the infection time-based algorithm on some datasets, especially real-world datasets. (3) Comparing the performance of algorithms on synthetic datasets, even with the same network size, different network properties can lead to drastically different results.

Table 3. Diffusion network inference results (F1-score). (The first three methods are infection time-based methods and the last two methods are infection time-independent methods.)

Alg.	Datasets					
	Random	Hierarchical	Core-periphery	Dolphin	Polblogs	Facebook
REFINE	0.316	0.315	0.270	0.520	0.372	0.289
InfoPath	0.796	0.799	0.778	0.752	0.634	0.403
NetINF	0.852	0.929	0.740	0.794	0.683	0.427
DeepINFER	0.464	0.499	0.381	0.792	0.574	0.335
InDNI	0.586	0.547	0.469	0.811	0.611	0.394
	+26.3%	+17.4%	+23.1%	+2.39%	+6.45%	+17.6%

Sensitivity Analysis Experiments. In this subsection, we explore the effects of the number of nodes, the number of edges, the number of the historical propagation and the hyperparameter ρ on the algorithm, respectively, as shown in Fig. 1. All experiments are performed on synthetic datasets. We control the Kronecker model to generate Hierarchical networks with uniform network mechanisms but with different parameters. From the figure we can observe that (1) The infection time-independent algorithms are more sensitive to network size than the infection time-based algorithm. (2) Compared with the infection time-independent algorithms, the infection time-based algorithm can obtain the set of potential parent nodes through the node infection time, which helps them achieve better results. (3) When the number of propagation is small, the statistical effect of propagation cannot be fully reflected, and the results of each algorithm are poorer under the influence of biased data. However, InDNI, NetINF and InfoPath have lower requirements for the number of propagations, which is important for practical applications. (4) The loss weight ρ avoids focusing too much on the zero element, and the appropriate parameter make the model significantly improve on each dataset.

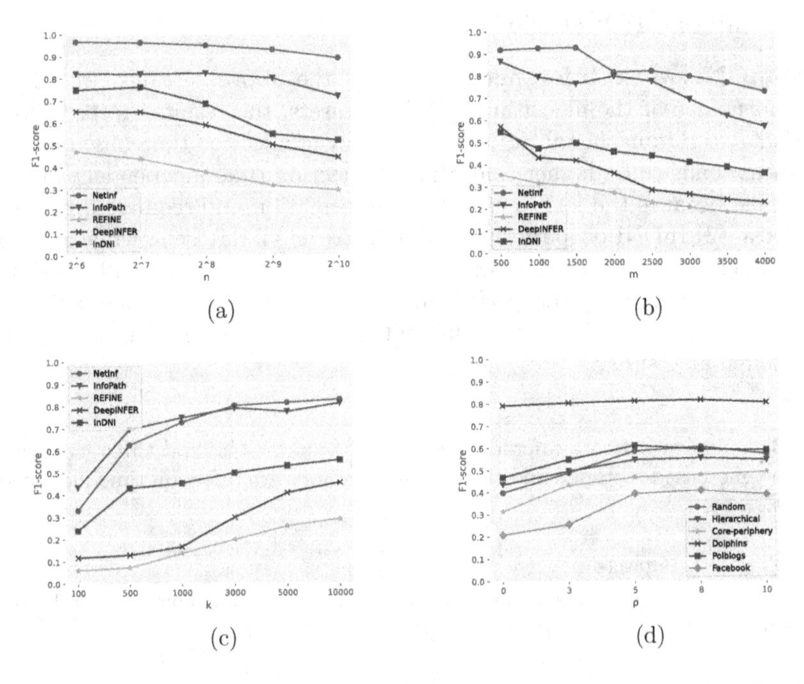

Fig. 1. Sensitivity analysis experiments. (a) Effects of the number of nodes; (b) Effects of the number of edges; (c) Effects of the number of propagation; (d) Effects of hyper-parameter ρ.

Ablation Experiment. To verify the effectiveness as well as the necessity of each part of the algorithm, we explored the performance of co-occurrence, MI, IMI, and InDNI without IMI versus InDNI, respectively, as shown in Fig. 2. From which we can observe that (1) The experimental results of IMI, InDNI without IMI and InDNI shows that filtering method, which plays an important role in the overall performance of the algorithm, does solve the problem of aggregation of isolated nodes in the embedding space. (2) The results of IMI are all better than MI, justifying the rationality and necessity of IMI. (3) The combination of node representation and filtering methods can achieve better results.

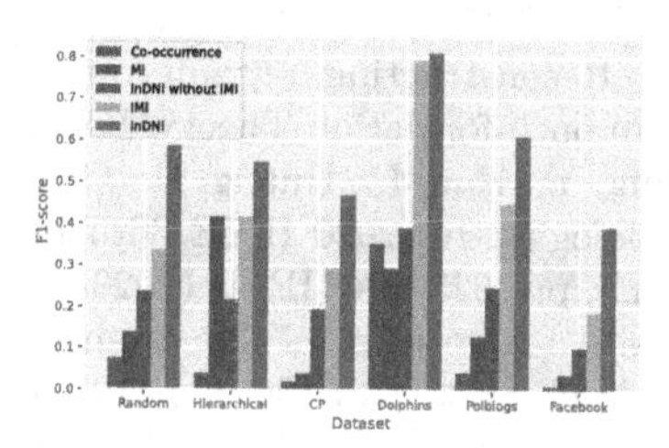

Fig. 2. Ablation experiment.

6 Conclusion and Future Work

In this paper, we propose an algorithm InDNI for diffusion network inference from the final infection status of nodes only. Compared with existing algorithms, InDNI does not depend on infection time or infection sequence and does not have strong a priori assumptions. Experiments on a large number of synthetic and real-world datasets demonstrate that InDNI outperforms other algorithms that do not rely on infection time, and is able to approach the performance of infection time based algorithms on some datasets.

In the future, we will subsequently consider data collection error detection and complementary algorithms for more precise prediction. In addition, the InDNI algorithm still has a gap with the infection time based algorithm. We believe that the learning of node representation can be improved and will explore it in future work.

Acknowledgements. This work was funded by the National Natural Science Foundation of China under grant numbers U1836111 and the National Social Science Fund of China under grant number 19ZDA329.

References

1. Adamic, L.A., Glance, N.: The political blogosphere and the 2004 us election: divided they blog. In: Proceedings of the 3rd International Workshop on Link Discovery, pp. 36–43 (2005)

2. Amin, K., Heidari, H., Kearns, M.: Learning from contagion (without timestamps). In: International Conference on Machine Learning. PMLR (2014)
3. Clauset, A., Moore, C., Newman, M.E.: Hierarchical structure and the prediction of missing links in networks. Nature **453**(7191), 98–101 (2008)
4. Doersch, C.: Tutorial on variational autoencoders. arXiv preprint arXiv:1606.05908 (2016)
5. Erdos, P., Rényi, A., et al.: On the evolution of random graphs. Publ. Math. Inst. Hung. Acad. Sci **5**(1), 17–60 (1960)
6. Givens, C.R., Shortt, R.M.: A class of Wasserstein metrics for probability distributions. Mich. Math. J. **31**(2), 231–240 (1984)
7. Gomez-Rodriguez, M., Leskovec, J., Krause, A.: Inferring networks of diffusion and influence. ACM Trans. Knowl. Discov. Data (TKDD) **5**(4), 1–37 (2012)
8. Gripon, V., Rabbat, M.: Reconstructing a graph from path traces. In: 2013 IEEE International Symposium on Information Theory. IEEE (2013)
9. Han, K., Tian, Y., Zhang, Y., Han, L., Huang, H., Gao, Y.: Statistical estimation of diffusion network topologies. In: 2020 IEEE 36th International Conference on Data Engineering (ICDE), pp. 625–636. IEEE (2020)
10. Hartigan, J.A., Wong, M.A.: Algorithm as 136: a k-means clustering algorithm. J. Roy. Stat. Soc. Ser. C (Appl. Stat.) **28**, 100–108 (1979)
11. Huang, H., Yan, Q., Gan, T., Niu, D., Lu, W., Gao, Y.: Learning diffusions without timestamps. In: Proceedings of the AAAI Conference on Artificial Intelligence, vol. 33, pp. 582–589 (2019)
12. Kefato, N.Z., Montresor, A.: DeepInfer: diffusion network inference through representation learning. In: Proceedings of the 13th International Workshop Mining Learning Graphs (2017)
13. Kefato, Z.T., Sheikh, N., Montresor, A.: REFINE: representation learning from diffusion events. In: Nicosia, G., Pardalos, P., Giuffrida, G., Umeton, R., Sciacca, V. (eds.) LOD 2018. LNCS, vol. 11331, pp. 141–153. Springer, Cham (2018). https://doi.org/10.1007/978-3-030-13709-0_12
14. Kipf, T.N., Welling, M.: Variational graph auto-encoders. arXiv preprint arXiv:1611.07308 (2016)
15. Kurashima, T., Iwata, T., Takaya, N., Sawada, H.: Probabilistic latent network visualization: inferring and embedding diffusion networks. In: Proceedings of the 20th ACM SIGKDD International Conference on Knowledge Discovery and Data Mining, pp. 1236–1245 (2014)
16. Leskovec, J., Chakrabarti, D., Kleinberg, J., Faloutsos, C., Ghahramani, Z.: Kronecker graphs: an approach to modeling networks. J. Mach. Learn. Res. **11**(2) (2010)
17. Leskovec, J., Lang, K.J., Dasgupta, A., Mahoney, M.W.: Statistical properties of community structure in large social and information networks. In: Proceedings of the 17th International Conference on World Wide Web, pp. 695–704 (2008)
18. Leskovec, J., Mcauley, J.: Learning to discover social circles in ego networks. Adv. Neural Inf. Process. Syst. **25** (2012)
19. Lusseau, D., Schneider, K., Boisseau, O.J., Haase, P., Slooten, E., Dawson, S.M.: The bottlenose dolphin community of doubtful sound features a large proportion of long-lasting associations. Behav. Ecol. Sociobiol. **54**(4), 396–405 (2003)
20. Mikolov, T., Sutskever, I., Chen, K., Corrado, G.S., Dean, J.: Distributed representations of words and phrases and their compositionality. Adv. Neural Inf. Process. Syst. **26** (2013)

21. Rodriguez, M.G., Leskovec, J., Balduzzi, D., Schölkopf, B.: Uncovering the structure and temporal dynamics of information propagation. Netw. Sci. **2**(1), 26–65 (2014)
22. Wang, H., Banerjee, A.: Bregman alternating direction method of multipliers. In: Advances in Neural Information Processing Systems, vol. 27. Curran Associates, Inc. (2014)
23. Xia, Y., Chen, T.H.Y., Kivelä, M.: Applicability of multilayer diffusion network inference to social media data. arXiv preprint arXiv:2111.06235 (2021)
24. Zhu, D., Cui, P., Wang, D., Zhu, W.: Deep variational network embedding in Wasserstein space. In: Proceedings of the 24th ACM SIGKDD International Conference on Knowledge Discovery & Data Mining (2018)

Beyond Precision: A Study on Recall of Initial Retrieval with Neural Representations

Yan Xiao, Yixing Fan(✉), Ruqing Zhang, and Jiafeng Guo

CAS Key Lab of Network Data Science and Technology, Institute of Computing Technology, Chinese Academy of Sciences, Beijing 100190, China
{fanyixing,zhangruqing,guojiafeng}@ict.ac.cn

Abstract. Vocabulary mismatch is a central problem in information retrieval (IR), i.e., the relevant documents may not contain the same (symbolic) terms of the query. Recently, neural representations have shown great success in capturing semantic relatedness, leading to new possibilities to alleviate the vocabulary mismatch problem in IR. However, most existing efforts in this direction have been devoted to the re-ranking stage. That is to leverage neural representations to help re-rank a set of candidate documents, which are typically obtained from an initial retrieval stage based on some symbolic index and search scheme (e.g., BM25 over the inverted index). This naturally raises a question: if the relevant documents have not been found in the initial retrieval stage due to vocabulary mismatch, there would be no chance to re-rank them to the top positions later. Therefore, in this paper, we study the problem how to employ neural representations to improve the recall of relevant documents in the initial retrieval stage. Specifically, to meet the efficiency requirement of the initial stage, we introduce a neural index for the neural representations of documents, and propose two hybrid search schemes based on both neural and symbolic indices, namely the parallel search scheme and the sequential search scheme. Our experiments show that both hybrid index and search schemes can improve the recall of the initial retrieval stage with small overhead.

Keywords: Indexing · Neural representation · Initial retrieval

1 Introduction

IR pipeline in modern search systems typically consists of two stages [7], namely the initial retrieval stage and the re-ranking stage. The initial retrieval stage aims to retrieve a small subset from the whole corpus that contains as many relevant documents as possible (i.e., high recall) with small cost (i.e., high efficiency). Without loss of generality, this is usually achieved under a symbolic index and search scheme in modern search systems. The re-ranking stage, on the other hand, aims to produce a high-quality (i.e., high precision) ranking list of the subset. Since the subset usually contains much fewer documents than the

Y. Chang and X. Zhu (Eds.): CCIR 2022, LNCS 13819, pp. 76–89, 2023.
https://doi.org/10.1007/978-3-031-24755-2_7

whole corpus, more complicated ranking algorithms, such as learning to rank algorithms [10,18] or deep neural models [12,13,15], could be involved in this stage for the re-ranking task.

The above pipeline has been widely adopted in most practical search systems, and a long-standing challenge it faces is the vocabulary mismatch problem, i.e., the relevant documents may not contain the same (symbolic) terms of the query. While there have been many efforts in developing ranking algorithms to address this challenge [14,15,28], most of them were at the re-ranking stage. But what if the relevant documents have not been found in the initial retrieval stage due to vocabulary mismatch, which is very likely to happen due to the symbolic index and search scheme? In this case, there would be no chance to re-rank those missing relevant documents to the top positions later. Therefore, we argue that it is critical to tackle the challenge at the very beginning. In other words, we shall attempt to solve the vocabulary mismatch problem at the initial retrieval stage, rather than only addressing it at the re-ranking stage.

In recent literature, neural representation (e.g., word embedding [23]) has achieved great success in capturing the semantic relatedness. By representing each word as a dense vector, similar words would be close to each other in the semantic space and the linguistic relations between words could be simply calculated via algebra. Such neural representations bring new possibility to alleviate the vocabulary mismatch problem in IR beyond the traditional symbolic term-based representation. Unfortunately, most existing efforts [15] in using neural representations for IR have been devoted to the re-ranking stage.

In this paper, we study the problem how to employ neural representations to improve the recall of relevant documents in the initial retrieval stage. To address this problem, we need to solve two major challenges, i.e., how to index neural representation of documents and how to search with neural index. To meet the efficiency requirement of the initial stage, we represent each document as a weighted sum of word embeddings, and introduce a k-nearest-neighbor (k-NN) graph based neural index which is efficient in both construction and search over dense vectors. We then propose two hybrid search schemes based on both neural and symbolic indices, namely the parallel search scheme and the sequential search scheme. The parallel search scheme retrieves documents based on symbolic index and neural index simultaneously, and merges the top results together to obtain the candidate subset. In this way, both the symbolic and neural indices act as two separate memories of the corpus. In the sequential search scheme, we first retrieve seed documents based on the symbolic index, and then expand the candidate subset based on the neural index. To evaluate the effectiveness and efficiency of our proposed hybrid index and search schemes, we conduct extensive experiments on two IR benchmark collections. The experiments show that by using neural index and hybrid search scheme, we could improve recall with small overhead for initial retrieval.

Overall, the major contributions of our work are as follows:

- We propose to enhance recall at the initial retrieval stage with neural representations. We introduce a k-NN graph based neural index and further

propose two hybrid search schemes, i.e., the parallel search scheme and the sequential search scheme.
- We conduct extensive experiments on two IR benchmark collections to evaluate the effectiveness and efficiency of our proposed approaches for initial retrieval.
- We conduct detailed analysis to study the utility and difference of both symbolic and neural indices.

The rest of the paper is organized as follows. Section 2 gives a brief summary of related work. We describe the detailed implementation of hybrid index and search schemes in Sect. 3. In Sect. 4 we present the experimental results and conduct detailed analysis. Section 5 concludes the paper and talks about the future work.

2 Related Work

In this section, we introduce the related work, including the existing methods for initial retrieval and explorations of neural representations for IR.

2.1 Initial Retrieval

Conventional initial retrieval relies on an inverted index to obtain a list of document candidates, and then simple model such as BM25 can be fast executed over these candidates to retrieve initial results. Two-stage learning to rank [7] was proposed to replace BM25 with a more complex ranker. This ranker is learned beyond query terms, including weighted phrases, proximities, and expansion terms. However, this has not alleviated the vocabulary mismatch problem since candidates are still obtained from the inverted index, and the efficiency has not been taken into consideration in this work either.

Other efforts have attempted to replace the symbolic inverted index with k-NN search. [20] proposed the two-stage hashing scheme for fast document retrieval. In their work, they represent both query and document as TF-IDF weight vectors, and use the cosine similarity to evaluate the similarity between query and document. They perform k-NN search using Locality Sensitive Hashing (LSH) [8] to retrieve document candidates in the first stage and then re-rank these documents using Iterative Quantization [11]. The two-stage hashing scheme can be more efficient than traditional IR baselines, but has not achieved the same effectiveness. Meanwhile, they evaluated by the precision while not by the recall.

To address the vocabulary mismatch problem, [2] proposed to use complex initial retrieval model and perform k-NN search in non-metric space. They use the linear combination of BM25 and IBM Model I [4] as the non-metric similarity, and find pivot-based index Neighborhood Approximation (NAPP) [27] can achieve some good results on question answering (QA) datasets. NAPP selects several documents as the pivots, and each query and document is represented by its k nearest pivots (k-NPs) computed by the brute force search. Given a query, the documents sharing a pre-specified number of k-NPs with the query are filtered to compute real distance. However, k-NN search in non-metric space leads to new indexing challenges. On the other hand, the effectiveness and efficiency of non-metric k-NN search have not been evaluated on the initial retrieval for IR, which is different with QA. Note that they have performed k-NN search using cosine similarity between averaged word embeddings of questions and answers, but this can not achieve good effectiveness.

2.2 Neural Representations for IR

Existing work of neural representations for IR mainly explore two ways, i.e., leveraging word embeddings to enhance the representations of query and document [13,24], or learning query and document representations by a deep model [15]. [13] proposed to build local interactions between each pair of words from a query and a document based on word embeddings. In their work, the local interactions are mapped into a fixed-length matching histogram, and then this histogram serves as the input of a deep model to learn the relevance matching score. [24] proposed a new ranking method based on comparing each query word to a centroid of the document word embeddings. In their work, the word embeddings used for query and document are in dual embedding space. Their model is effective in re-ranking top documents returned by a search engine, and a linear mixture of their model and BM25 can be employed to rank a larger set of candidate document. Rather than leveraging word embeddings directly, [17] proposed to learn query and document representations from clickthrough logs by a deep model and model the relevance by cosine similarity between query and document representations. All these work have shown that employing neural representations for IR can achieve better effectiveness. However, these work aims at the re-ranking stage, while not the initial retrieval stage. If a relevant document is not contained in the initial results, the re-ranking stage can only generate a sub-optimal ranking list.

3 Our Approach

In this section, we firstly introduce the symbolic index and neural index used in our approach, and then describe the parallel and sequential search schemes based on these two kinds of indices. Finally, we give some discussions on these two search schemes.

3.1 Symbolic Index

Conventional IR is based on symbolic term-based representation, where query and document are represented as a bag of terms (i.e., Bag-of-Words representation), and each term is represented as a one-hot vector. Each dimension of such a symbolic representation denotes the occurrence times of one distinct term, and is treated as independent from others. Due to the sparsity of symbolic representation, inverted index is widely employed as the core index structure in model search systems. In an inverted index, each term is linked to a posting list of its occurrence information in the corpus, including the document identifier, corresponding frequency and so on.

Given a corpus, the documents are scanned one by one to obtain their symbolic representations and then it is efficient to construct an inverted index. Specifically, for each non-zero dimension of a symbolic representation, the occurrence information is appended to the posting list of the corresponding term.

3.2 Neural Index

In this work, we further employ neural representation for documents so that we can search documents beyond traditional symbolic terms to enhance recall at the initial retrieval stage. Rather than learning a representation for each document, we simply adopted the TF-IDF weighted sum of word embeddings as its neural representation due to the trade-off between effectiveness and computation efficiency [2,24]. Since the neural representations are continuous and dense vectors, the inverted index is not suitable for searching over them.

Indexing technologies for dense vectors have been widely studied and many structures have been proposed, such as k-d tree [1], LSH [8], neighborhood graph-based methods [16,22]. Here we adopt a k-nearest neighbor (k-NN) graph [16] based structure as the neural index. Each document is linked to its k most semantically similar documents measured by cosine similarity between their neural representations. The links can be reversed to obtain undirected k-NN graph [21], i.e., attaching the reversed neighbors behind the k-NNs for each document.

Since constructing a precise k-NN graph is time-consuming for a large scale dataset, many approximate algorithms [6,9] have been proposed. We adopt the state-of-the-art algorithm NN-Descent [9], which can be extremely efficient to construct a highly precise k-NN graph. In an evolving dataset, the neighbors will change with insertions. Rather than rebuilding the whole graph, we could perform approximate nearest neighbor search to find the k-NNs for a new inserted point and update the neighbors of these k-NNs with the new inserted point.

3.3 Parallel Search Scheme

With both the symbolic and neural indices in hand, a natural idea is to retrieve based on these two kinds of indices simultaneously, and merge the top results together to obtain the candidate subset. In this way, both symbolic and neural indices act as two separate memories of the corpus. As there are two search paths, we name this scheme as the parallel search scheme (ParSearch in short).

The symbolic search is conducted in the left path, where posting lists of the corresponding query terms are first looked up through the inverted index and merged to obtain the candidate documents. Then we use BM25 to retrieve m candidates from these documents, where m is pre-specified. Simultaneously, in the right path, the neural search is performed to find n nearest neighbors of the query in the semantic space via traversing on the graph index [16], where n is the number of candidate documents for initial retrieval. Specifically, some documents are sampled as starting points, and the linked documents of the most similar and unexplored document are iteratively explored to approach the query. At each iteration, n most similar documents to the query are kept. The similarity between query and document is measured by cosine similarity between their neural representations.

The results of symbolic search (i.e., BM25) are based on exact matching signals, while the results of neural search (i.e., cosine similarity) are based on semantic matching signals. Since these two kinds of documents are not scored in the same space and symbolic search is more precise than neural search as shown in previous work [2,24], thus we introduce a new aggregation method to get final results. Specifically, we take the set of documents from the symbolic search as the base, and merge those from the neural search into it to obtain sufficient number of candidate documents. In the merging process, we scan the documents from neural search from top to end one by one, and if it is not in the symbolic candidate set, it will be merged. This merging process is efficient since we do not need to compute any score or re-rank any document. This is reasonable as we only care about the recall rather than the precision.

The parallel search scheme is formally described in Algorithm 1.

Algorithm 1. Parallel Search Scheme

Input: query q, number of initial candidate documents n, number of candidate documents retrieved by symbolic search m

Output: n initial candidate documents

```
M ← m documents based on the inverted index and BM25
C ← n documents based on the undirected k-NN graph index and cosine similarity
for each d ∈ C do
    if d ∈ M then
        continue
    end if
    M ← M ∪ {d}
    if |M| == n then
        break
    end if
end for
return M
```

3.4 Sequential Search Scheme

Although parallel search scheme is straightforward and efficient, it may not be very effective since the retrieval performance of neural representation alone is questionable. Another possible way to combine symbolic and neural search is a sequential manner, with the expectation that documents semantically similar to relevant documents are expected to be relevant. In this way, we can first employ symbolic search to find some candidate documents as the seeds, and then use neural search to expand the seeds to obtain the final candidate subset. In this case, the symbolic index acts as the precise memory while the neural index acts as the associative memory of the corpus. As there is only one chained search path, we name this scheme as the sequential search scheme (SeqSearch in short).

Given a query, we firstly conduct symbolic search to retrieve seed documents as the same with ParSearch. We expand these seeds to associate their semantically similar documents based on the graph index, i.e., obtaining their k-NNs via looking up the graph. Then cosine similarity is used to obtain the top associated documents. We return the seeds with the later retrieved candidates together as the final results.

To control the efficiency overhead from neural search, we further propose a heuristic strategy to reduce the computation, i.e., only expanding a pre-specified proportion of highly scored seed documents. The underlying intuition is that k-NNs of a seed are more relevant if the seed has higher exact matching score. We do not reduce the number of seeds since exact matching is the most important signal in IR [13] and more reliable than semantic matching.

Algorithm 2. Sequential Search Scheme

Input: query q, k-NN graph G, number of initial candidate documents n, number of seed documents s, expanding proportion p
Output: n initial candidate documents
$S \leftarrow s$ seed documents based on the inverted index and BM25
$D \leftarrow \emptyset$
for each $d \in S[1 : p * s]$ **do**
  $U \leftarrow k$ neighbors of d in G
  $D \leftarrow D \cup \{U \setminus S\}$
end for
$N \leftarrow$ Retrieve $n - s$ documents from D based on cosine similarity
return $S \cup N$

3.5 Discussions

The ParSearch and SeqSearch both employ neural representations to improve the recall of relevant documents in the initial retrieval stage. Here we make detailed discussions on them.

Efficiency. The ParSearch considers symbolic search and neural search separately, which can be efficient since these two processes can be conducted at the same time. In our experiments, we find that the neural search executes more efficiently than symbolic search. Thus the ParSearch only has small and almost constant overhead from the merging process.

As for the SeqSearch, neural search is executed after symbolic search, thus may sacrifice some efficiency. Note here in SeqSearch, we only need to look up k-NNs for each seed document, where the computational complexity is O(1).

Therefore, the efficiency overhead of neural search is almost constant and has no relationship with the query length and collection size. With the increasing of query length or corpus size, an increasing number of documents are likely to contain the terms of the query, leading to more cost of symbolic search (i.e., more documents are initially retrieved by inverted index and scored by BM25). The relative overhead from neural search would become negligible as compared with the symbolic search.

Effectiveness. Although it is more efficient to retrieve separately as ParSearch does, this may have a disadvantage against the effectiveness since semantic matching with neural representation alone may bring many noise matching signals [25], e.g., any two adjective terms could contribute some matching signals.

Comparing with the ParSearch, neural search of SeqSearch is based on symbolic search, i.e., the semantically matched document are retrieved from the documents associated by exactly matched seeds. The association process could filter plenty of noise documents and ensure the quality of neural search. Our experiments have also demonstrated this.

Extendibility. The k-NN graph is organized by linking each document to its k most similar documents. We can define the similarity as a more powerful function, such as the linear combination of a exact matching model and a semantic matching model. Such a similarity function can be neither metric nor symmetric, but could be supported by NN-Descent [9].

Since performing k-NN search with non-metric graph is more complex and challenging [2] in ParSearch than association in SeqSearch, it seems SeqSearch has better extendibility than ParSearch. The major reason lies in that the ParSearch relies on the neural index to find k-NNs for unseen queries, while the SeqSearch only need a neural index to look up k-NNs for existing documents.

4 Experiments

In this section, we conduct experiments to evaluate the effectiveness and efficiency of ParSearch and SeqSearch on two IR benchmark collections. We conduct experiments on two TREC collections, Robust04 and WT2G. Robust04 is a news dataset and its topics are collected from TREC Robust Track 2004. WT2G is a general Web crawl and its topics are collected from TREC Web Track 1999. We make use of both the title and the description of each TREC topic in our experiments. Both documents and queries are white-space tokenized, lowercased, and stemmed using the Krovetz stemmer [19]. Stop word removal is performed on query words during retrieval using the INQUERY stop list [5].

4.1 Baselines and Experimental Settings

We adopt the symbolic index based method and neural index based method as the baselines. These methods include:

BM25: The BM25 formula [26] is a highly effective symbol based initial retrieval model. We use the inverted index implemented in Apache Lucene[1]. We tune the parameters $k1$ and b to obtain the bast performance.

Cosine: The matching score is the cosine similarity between neural representations of query and document [3,24]. For fast initial retrieval, approximate k-NN search is performed as in [2]. We adopt undirected 20-NN graph [16,21] as the index, and tune the number of kept nearest candidates to control the trade-off. We use subscript to denote this number.

LinComb: The matching score is the linear combination of BM25 score and cosine similarity. Given the query q and a document d, the matching score is defined as below,

$$Matching\ Score(q, d) = \lambda * BM25(q, d) + (1 - \lambda) * Cosine(\vec{q}, \vec{d})$$

where λ denotes the co-efficiency balancing two scores. To choose the λ, we perform a parameter sweep between 0.0 and 1.0 at intervals of 0.01 using brute force.

ParSearch: For our proposed parallel search scheme, we adopt the same index as Cosine for neural representations.

SeqSearch: As for our proposed sequential search scheme, we use the same 20-NN graph in Cosine. We use subscript to denote the expanding proportion p.

The word embeddings used for neural representations are trained on Robust04 and WT2G collections respectively by the Skip-Gram model implemented in Word2Vec[2]. Specifically, we set *min-count* = 0 to keep all the words and use 200-dimension embeddings.

The experiments on Robust04 and WT2G are conducted on a machine with 2.7 GHz Intel Core i5-5257U CPU and 8 GB memory, and a single thread is used to test the performance for all methods. The neural indices of Cosine, ParSearchWe and SeqSearch are based on the same 20-NN graph. Constructing the graph costs 644s and 334s on Robust04 and WT2G respectively using 4 threads.

4.2 Evaluation Methodology

To evaluate the effectiveness and efficiency of different index and search schemes for initial retrieval, we use averaged recall@1000 and time as the metrics. Since queries have different numbers of relevant document, we also report the ratio of all retrieved relevant documents. Given the limited number of queries for each collection, we randomly divide them into 5 folds and conduct 5-fold cross-validation.

4.3 Retrieval Performance and Analysis

This section presents the performance of different index and search schemes on two benchmark TREC collections, and the summary of results is displayed in

[1] http://lucene.apache.org.
[2] https://code.google.com/p/word2vec/.

Table 1. Comparison of different index and search schemes. Significant improvement or degradation with respect to BM25 for recall@1000 is indicated $(+/-)$(p-$value \leq 0.05$).

Index type	Index method	Search method	Topic titles			Topic descriptions		
			Recall (%)	Ratio (%)	Time (ms)	Recall (%)	Ratio (%)	Time (ms)
Robust04 collection								
Symbolic	Inverted Index	BM25	68.36	57.84	4.50	66.68	53.21	10.34
Neural	Graph index	Cosine_{100}	47.76^-	40.80	1.01	49.70^-	39.83	1.17
		Cosine_{500}	49.89^-	42.64	2.64	50.14^-	39.61	2.99
		Cosine_{800}	50.27^-	42.67	3.54	50.14^-	39.47	4.04
		Cosine_{1000}	50.37^-	42.80	4.17	50.31^-	39.57	4.73
Neural and symbolic	NAPP	LinComb_5	62.13^-	53.98	163.59	62.51^-	51.48	212.40
		LinComb_{10}	28.15^-	27.27	59.15	30.35^-	27.89	88.61
		LinComb_{15}	16.23^-	16.21	28.15	17.01^-	16.30	42.35
	Inverted index and graph index	ParSearch	72.06^+	61.45	5.14	68.46^+	55.28	10.58
		$\text{SeqSearch}_{25\%}$	72.34^+	62.05	5.45	69.15^+	55.92	11.86
		$\text{SeqSearch}_{50\%}$	72.02^+	61.80	6.37	69.02^+	55.87	13.06
		$\text{SeqSearch}_{100\%}$	71.87^+	61.76	8.30	68.83^+	55.84	15.20
WT2G collection								
Symbolic	Inverted index	BM25	82.03	82.23	6.38	78.81	77.62	11.90
Neural	Graph index	Cosine_{100}	44.19^-	42.21	0.84	45.76^-	43.62	0.96
		Cosine_{500}	50.04^-	52.17	2.20	51.11^-	49.98	2.36
		Cosine_{800}	50.65^-	52.70	2.98	51.29^-	50.11	3.16
		Cosine_{1000}	50.45^-	52.52	3.84	51.89^-	50.55	3.54
Neural and symbolic	NAPP	LinComb_5	69.07^-	70.34	116.43	71.04^-	71.92	138.06
		LinComb_{10}	18.13^-	19.31	44.09	20.59^-	23.08	56.69
		LinComb_{15}	3.83^-	5.05	16.23	4.98^-	6.41	20.90
	Inverted index and graph index	ParSearch	83.70	84.07	7.84	79.95^+	79.03	12.30
		$\text{SeqSearch}_{25\%}$	83.56	84.12	8.30	80.11^+	79.38	13.32
		$\text{SeqSearch}_{50\%}$	83.58	84.07	9.14	80.12^+	79.42	14.42
		$\text{SeqSearch}_{100\%}$	83.88	84.29	11.02	80.20^+	79.51	16.10

Table 1. Here, we adopt Cosine_{1000} for ParSearch since Cosine_{1000} performs the most precise k-NN search with a lower efficiency cost than BM25.

While comparing the performance of the recall@1000 metric, we can see that the neural index based method Cosine performs significantly worse than the symbolic index based method BM25 on two collections. This result suggests that exact matching signal plays an important role in IR, and neural representations could not well capture this signal. Meanwhile, ParSearch and SeqSearch both can achieve better effectiveness consistently, demonstrating that neural representations can be helpful to improve the recall of relevant documents. We find SeqSearch can achieve better performance than ParSearch consistently. The reason is that semantic matching with neural representation alone brings many noise matching signals as we have discussed in Sect. 3.5.

Although k-NN search in non-metric space is an interesting idea, we find LinComb performs extremely worse on the IR benchmark collections. [2] point out that NAPP is effective only if comparing a query and a document with the same pivot provides meaningful proximity information. In our IR experiments, the queries are extremely short. For example, the title and description of Robust04 are 2.63 and 8.16 on average correspondingly. The short query makes it hard to have overlap terms with limited pivots, leading to less effectiveness of BM25. Meanwhile, the cosine similarity is too coarse and comparing the query and its relevant documents with a nearly random pivot in the semantic space provides confusing information.

As for the efficiency, we find that Cosine can achieve the best performance, while ParSearch and SeqSearch cost more time than BM25 due to the additional process. Considering the improvement of the effectiveness, the efficiency overhead could be acceptable. We can see the relative additional cost of our search schemes both become smaller as the query length increases. For example, the title and description of Robust04 are 2.63 and 8.16 on average, while the additional cost of ParSearch are 14.22% and 2.32% of BM25 correspondingly. The reason is that longer descriptions lead to more documents being computed in symbolic search based on the inverted index, while our additional cost keep almost constant as the query length increases. Meanwhile, Cosine could cost almost the same time, since the query is always the 200-dimension vectors no matter what the original query length is.

Surprisingly, we find LinComb can not achieve good efficiency even at low recall. The reason is also due to the too short query, which leads to a lot of ineffective computations as many shared k-NPs are meaningless. In contrast, [2] retrieve the best answer using the question summary concatenated with the description. They report that k-NN search based on NAPP can be more than 1.5x faster than Lucene on Stack Overflow dataset[3] whose query length is 48.4 on average. But on Yahoo!Answers Comprehensive dataset[4] whose query length is only 17.8, their approach is 2x slower than Lucene when achieving about the same recall.

4.4 Analysis on Retrieved Relevant Documents

In this section, we take the retrieval using topic titles as example, and conduct detailed analysis on retrieved relevant documents to study the utility and difference of both symbolic and neural indices.

Table 2. Statistics of the averaged relevant documents retrieved by BM25 and Cosine_{1000} using topic titles.

	Robust04	WT2G
Intersection of BM25 and Cosine_{1000}	21.81	21.42
Only by BM25	18.47	16.06
Only by Cosine_{1000}	8.00	2.52

[3] The dump from https://archive.org/download/stackexchange dated March 10th 2016.

[4] https://webscope.sandbox.yahoo.com.

Since the results of ParSearch are from both BM25 and Cosine_{1000}, we analyze the relevant documents retrieved by BM25 and Cosine_{1000} separately, and the statistics are shown in Table 2. As we can see, more than half of the relevant documents by BM25 and Cosine_{1000} are the same, while others can only be retrieved by BM25 or by Cosine_{1000}. For example, on Robust04, 18.47 documents on average found by BM25 are not in the results of Cosine_{1000}. Meanwhile, Cosine_{1000} can find 8.00 relevant documents missed by BM25. Note that the relevant documents retrieved only by Cosine_{1000} is not too much on WT2G, thus leading to less improvement on WT2G. We can find that BM25 could find more relevant documents than Cosine_{1000}, this suggests the importance of exact matching.

Table 3. Statistics of the averaged relevant documents retrieved by seed and associated documents using topic titles.

	Robust04	WT2G
Seed documents	680	800
Relevant seed documents	36.91	36.44
Associated documents with 25% seeds	1773	1577
Relevant associated documents	10.28	3.76
Associated documents with 50% seeds	3441	2977
Relevant associated documents	12.80	4.60
Associated documents with 100% seeds	6666	5534
Relevant associated documents	15.80	5.58

As for the SeqSearch, the number of relevant documents contained in seed and associated documents are shown in Table 3. As we can see, the k-NNs in the semantic space of well exactly matched documents contains many relevant documents, e.g., 10.28 relevant documents in the 20-NNs of $700 * 25\% = 175$ seeds on Robust04. Note that we remove the seed documents from the initially associated documents, so the number of relevant documents may be larger in the 20-NNs of seeds. We can see that on WT2G, the proportion of relevant documents found in associated documents is not as big as on Robust04. Since exact matching has worked well enough on WT2G, the improvement on WT2G is thus not significant. Moreover, using a high proportion of seeds to associate can find a little more relevant documents while the number of associated documents increases exponentially, leading to more difficulty to distinguish the relevant documents with cosine similarity. This is the reason why $\text{SeqSearch}_{25\%}$ can achieve comparable performance with $\text{SeqSearch}_{100\%}$.

5 Conclusions

In this paper, we argue that neural representations can also be employed to improve the recall of relevant documents for initial retrieval. To solve the index and search challenges, we introduce a k-NN graph based neural index and further propose the parallel search scheme and the sequential search scheme based on both neural and symbolic indices. Our experiments show that both hybrid index and search schemes can improve the recall of the initial retrieval stage with small overhead. For future work, we would like to conduct empirical study of complex similarity in k-NN graph which may achieve better effectiveness.

Acknowledgement. This work was funded by the National Natural Science Foundation of China (NSFC) under Grants No. 61902381, No. 62006218, and No. 61732008, the Youth Innovation Promotion Association CAS under Grants No. 2021100, and 20144310, the Young Elite Scientist Sponsorship Program by CAST under Grants No. YESS20200121, the Lenovo- CAS Joint Lab Youth Scientist Project.

References

1. Bentley, J.L.: Multidimensional binary search trees used for associative searching. Commun. ACM **18**(9), 509–517 (1975)
2. Boytsov, L., Novak, D., Malkov, Y., Nyberg, E.: Off the beaten path: let's replace term-based retrieval with k-NN search. In: Proceedings of the 25th ACM International on Conference on Information and Knowledge Management, pp. 1099–1108. ACM (2016)
3. Brokos, G.-I., Malakasiotis, P., Androutsopoulos, I.: Using centroids of word embeddings and word mover's distance for biomedical document retrieval in question answering. arXiv preprint arXiv:1608.03905 (2016)
4. Brown, P.F., Pietra, V.J.D., Pietra, S.A.D., Mercer, R.L.: The mathematics of statistical machine translation: parameter estimation. Comput. Linguist. **19**(2), 263–311 (1993)
5. Callan, J.P., Croft, W.B., Broglio, J.: TREC and TIPSTER experiments with inquery. Inf. Process. Manag. **31**(3), 327–343 (1995)
6. Chen, J., Fang, H.-R., Saad, Y.: Fast approximate kNN graph construction for high dimensional data via recursive Lanczos bisection. J. Mach. Learn. Res. **10**(Sep), 1989–2012 (2009)
7. Dang, V., Bendersky, M., Croft, W.B.: Two-stage learning to rank for information retrieval. In: Serdyukov, P., et al. (eds.) ECIR 2013. LNCS, vol. 7814, pp. 423–434. Springer, Heidelberg (2013). https://doi.org/10.1007/978-3-642-36973-5_36
8. Datar, M., Immorlica, N., Indyk, P., Mirrokni, V.S.: Locality-sensitive hashing scheme based on P-stable distributions. In: Proceedings of the Twentieth Annual Symposium on Computational Geometry, pp. 253–262. ACM (2004)
9. Dong, W., Moses, C., Li, K.: Efficient k-nearest neighbor graph construction for generic similarity measures. In: Proceedings of the 20th International Conference on World Wide Web, pages 577–586. ACM (2011)
10. Freund, Y., Iyer, R., Schapire, R.E., Singer, Y.: An efficient boosting algorithm for combining preferences. J. Mach. Learn. Res. **4**(Nov), 933–969 (2003)
11. Gong, Y., Lazebnik, S., Gordo, A., Perronnin, F.: Iterative quantization: a procrustean approach to learning binary codes for large-scale image retrieval. IEEE Trans. Pattern Anal. Mach. Intell. **35**(12), 2916–2929 (2013)
12. Guo, J., Cai, Y., Fan, Y., Sun, F., Zhang, R., Cheng, X.: Semantic models for the first-stage retrieval: a comprehensive review. ACM Trans. Inf. Syst. **40**(4), 1–42 (2022)

13. Guo, J., Fan, Y., Ai, Q., Croft, W.B.: A deep relevance matching model for ad-hoc retrieval. In: Proceedings of the 25th ACM International on Conference on Information and Knowledge Management, pp. 55–64. ACM (2016)
14. Guo, J., Fan, Y., Ai, Q., Croft, W.B.: Semantic matching by non-linear word transportation for information retrieval. In: Proceedings of the 25th ACM International on Conference on Information and Knowledge Management, pp. 701–710. ACM (2016)
15. Guo, J., et al.: A deep look into neural ranking models for information retrieval. Inf. Process. Manag. **57**(6), 102067 (2020)
16. Hajebi, K., Abbasi-Yadkori, Y., Shahbazi, H., Zhang, H.: Fast approximate nearest-neighbor search with k-nearest neighbor graph. In: Proceedings-International Joint Conference on Artificial Intelligence, vol. 22, p. 1312 (2011)
17. Huang, P.-S., He, X., Gao, J., Deng, L., Acero, A., Heck, L.: Learning deep structured semantic models for web search using clickthrough data. In: Proceedings of the 22nd ACM International Conference on Conference on Information & Knowledge Management, pp. 2333–2338. ACM (2013)
18. Joachims, T.: Optimizing search engines using clickthrough data. In: Proceedings of the eighth ACM SIGKDD International Conference on Knowledge Discovery and Data Mining, pp. 133–142. ACM (2002)
19. Krovetz, R.: Viewing morphology as an inference process. In: Proceedings of the 16th Annual International ACM SIGIR Conference on Research and Development in Information Retrieval, pp. 191–202. ACM (1993)
20. Li, H., Liu, W., Ji, H.: Two-stage hashing for fast document retrieval. In: Proceedings of the 52nd Annual Meeting of the Association for Computational Linguistics (Volume 2: Short Papers), vol. 2, pp. 495–500 (2014)
21. Li, W., Zhang, Y., Sun, Y., Wang, W., Zhang, W., Lin, X.: Approximate nearest neighbor search on high dimensional data–experiments, analyses, and improvement (v1. 0). arXiv preprint arXiv:1610.02455 (2016)
22. Malkov, Y.A., Yashunin, D.: Efficient and robust approximate nearest neighbor search using hierarchical navigable small world graphs. arXiv preprint arXiv:1603.09320 (2016)
23. Mikolov, T., Sutskever, I., Chen, K., Corrado, G.S., Dean, J.: Distributed representations of words and phrases and their compositionality. In: Advances in Neural Information Processing Systems, pp. 3111–3119 (2013)
24. Mitra, B., Nalisnick, E., Craswell, N., Caruana, R.: A dual embedding space model for document ranking. arXiv preprint arXiv:1602.01137 (2016)
25. Pang, L., Lan, Y., Guo, J., Xu, J., Cheng, X.: A deep investigation of deep IR models. arXiv preprint arXiv:1707.07700 (2017)
26. Robertson, S.E., Walker, S.: Some simple effective approximations to the 2-poisson model for probabilistic weighted retrieval. In: Croft, B.W., van Rijsbergen, C.J. (eds.) SIGIR 1994, pp. 232–241. Springer, New York (1994). https://doi.org/10.1007/978-1-4471-2099-5_24
27. Tellez, E.S., Chavez, E., Navarro, G.: Succinct nearest neighbor search. Inf. Syst. **38**(7), 1019–1030 (2013)
28. Xu, J., Wu, W., Li, H., Xu, G.: A kernel approach to addressing term mismatch. In: Proceedings of the 20th International Conference Companion on World Wide Web, pp. 153–154. ACM (2011)

A Learnable Graph Convolutional Neural Network Model for Relation Extraction

Jinling Xu, Yanping Chen(✉), Yongbin Qin, and Ruizhang Huang

State Key Laboratory of Public Big Data, College of Computer Science and Technology Guizhou University, Guiyang 550025, China
ypench@gmail.com, {ybqin,rzhuang}@gzu.edu.cn

Abstract. Relation extraction is the task of extracting the semantic relationships between two named entities in a sentence. The task relies on semantic dependencies relevant to named entities. Recently, graph convolutional neural networks have shown great potential in supporting this task, wherein dependency trees are usually adopted to learn semantic dependencies between entities. However, the requirement of external toolkits to parse sentences poses a problem, owing to them being error prone. Furthermore, entity relations and parsing structures vary in semantic expressions. Therefore, manually designed rules are required to prune the structure of the dependency trees. This study proposed a novel learnable graph convolutional neural network model (L-GCN) that directly encodes every word of a sentence as nodes of a graph neural network. Then, the L-GCN uses a learnable adjacency matrix to encode dependencies between nodes. The model offers the advantage of automatically learning high-order abstract representations of the semantic dependencies between words. Moreover, a fusion module was designed to aggregate the global and local semantic structure information of sentences. Further, the proposed L-GCN was evaluated on the ACE 2005 English dataset and Chinese Literature Text Corpus. The experimental results confirmed the effectiveness of L-GCN in learning the semantic dependencies of a relation instance. Moreover, it clearly outperformed previous dependency-tree-based models.

Keywords: Relation extraction · Graph convolutional neural network · Multi-channel · nlp

1 Introduction

Relation extraction is the task of identifying predefined relationships between two entities within a sentence. For example, given a sentence "`the billionaire was sentenced to one year in prison`" and two entities "`the billionaire`" and "`prison`" within it, relation extraction identifies the relationship between them as "`PHYS (physical)`". This is a fundamental natural

J. Xu, Y. Chen, Y. Qin and R. Huang—These authors contributed equally to this work.

Y. Chang and X. Zhu (Eds.): CCIR 2022, LNCS 13819, pp. 90–104, 2023.
https://doi.org/10.1007/978-3-031-24755-2_8

language processing task and can be applied to support many downstream tasks such as summarization [1], text mining [2], and question answering [3]. In related works, relation extraction has usually been implemented as a classification problem to predict confidence scores for all class categories-based features relevant to an entity pair in a sentence. Because the types of relations between entities are asymmetrical, all entity pairs in a sentence must be verified, which results in a serious data imbalance problem. Furthermore, a sentence often contains several named entities, and all entity pairs share the same context. Therefore, the extraction of relationships is a challenging task.

In recent years, various neural network models have been applied to support relation extraction [4,5]. They offer the advantage of automatically learning high-order semantic features from raw inputs, thereby avoiding the requirement of depending on manually designed features. The main problem is that traditional models are directly implemented on a raw input, which is weak in capturing semantic dependencies between two entities in a relational instance. Because a relation is a semantic expression relevant to two named entities in a sentence, the manner in which the structural information of a sentence can be effectively captured is particularly important for the task of relation extraction. In related studies, the dependency tree has been widely used for learning semantic dependencies. For example, Yan et al. [6] applied the shortest path of the two entities in the dependency tree to the long- and short-term memory network (LSTM) for relation extraction tasks. Miwa and Bansal [7] conducted experiments to compare the performance of the shortest path subtree, least common ancestor (LCA) subtree, and entire tree of a dependency tree. Veyseh et al. [8] proposed an ON-LSTM model that combined a dependency tree to obtain the syntactic-based importance score of each word in a sentence.

Recently, many researchers have implemented graph neural networks on dependency trees to learn the semantic structure information of sentences because they facilitate encoding of node representations relevant to adjacent nodes. Many researchers use the dependency tree of sentences to build the graph structure required by graph neural networks. When the dependency tree was used for the first time, many researchers converted the entire dependency tree into a graph [9,10]. Later, it was discovered that the entire information in the entire dependency tree did not contribute to the relation extraction task. Therefore, certain researchers have used the LCA path of two entities in the dependency tree to transform it into a graph [11]. In addition, many changes have occurred in the process of transforming a dependency tree into a graph. Previously, the dependency tree was transformed into a fully connected graph without considering the direction information of the edge in the tree [12]. Later, while maintaining the direction information of the edge in the tree, syntactic dependency between nodes was considered [13]. Thus, irrelevant information in the dependency tree can be effectively ignored while exploiting the relevant information in the dependency tree.

However, this method of generating graphs using dependency trees has two drawbacks. First, owing to the process of parse tree generation being error prone, the error is propagated to the entire model, which affects the performance of the model. Second, because a dependency tree is generated for a specific sentence, training in batches during the training of the model is challenging, thereby resulting in complicated calculations.

Thus, this study proposed a novel learnable convolutional graph neural network model (L-GCN). Specifically, each word was regarded as a node, and the edges between them were randomly initialized and changed during the model training process. Subsequently, a graph neural network was used to extract the relationship between each word in the sentence and other words. Moreover, inspired by the multi-channel neural network model (which can capture the structural information of a sentence) [14], a sentence was divided into five parts based on two entities. Consequently, to better obtain the structural information of the sentence, the position information of each word in the sentence and a tag containing the entity type were added alongside the two entities. In addition, a fusion module was used to aggregate the global and local semantic structure information. The global and local semantic structure information were obtained via the graph convolutional neural network and convolutional neural network, respectively. Thus, the proposed model can learn more comprehensive semantic structure information from sentences. Moreover, better performance was achieved on both the ACE2005 English dataset and Chinese Literature Text Corpus.

The contributions of this study are as follows:

- A novel L-GCN model was proposed, wherein a learnable graph was constructed by considering each word in the sentence as a node, and using a learnable adjacency matrix to encode the dependency between nodes.
- A fusion module that can capture the global and local semantic structure information of sentences was designed to enable the model in learning more comprehensive semantic structure information.
- The empirical study on ACE2005 English dataset and the Chinese Literature Text Corpus confirmed the effectiveness of the proposed model.

The remainder of this paper is organized as follows. Section 2 presents the related work. Section 3 elaborates on the construction process of the proposed model. Further, details of the experiment are presented in Sect. 4. Finally, Sect. 5 presents the conclusions and future work.

2 Related Work

Feature-Based Models: The previous methods used hand-designed features for relation extraction tasks. Such models are referred to as feature-based models. These models primarily focus on obtaining better features. Rink and Harabagiu [15] achieved the best results for SemEval-2010 Task 8 by designing effective features. However, this feature-based model relied heavily on the quality of designed features.

Neural Network-Based Models: With the rise of neural networks, an increasing number of relation extraction models based on neural networks are being proposed. They are mainly divided into dependency- and sequence-based models. The former uses external tools to generate a parse tree for sentences containing entities and then applies the parse tree to the relation extraction model. The dependency tree has been proven to effectively capture long-distance dependencies. However, to use the dependency tree more effectively, in addition to using the dependency path in the dependency tree, various pruning strategies have been gradually developed. The primary reason for pruning the dependency tree is to prune out certain irrelevant information and exploit the information that is valuable to the model. In previous studies, the pruning strategy mainly included preserving the shortest and LCA dependency paths between two entities [7]. To generalize the pruning strategy to many examples, certain researchers have used the self-attention mechanism to transform the dependency tree into a fully connected weighted graph [12]. However, the sequence-based model [16,17] primarily focuses on the information of the sentence itself. Moreover, it primarily applies various neural network models to directly affect the sentence itself.

Recently, owing to the powerful expressive ability of graphs, graph convolutional neural networks (GCNs) have been successfully applied to relation extraction tasks [18]. Constructing an effective graph plays a vital role in the process of using GCNs. In one case, the graph is constructed with the help of a dependency tree. Fu and Ma [9] converted the complete dependency tree into a graph, whereas, Zhang et al. [11] transformed the LCA dependency paths of both entities in a dependency tree into a graph. In another case, the graph is constructed directly using the information of the sentence itself. Vashishth et al. [19] regarded entities and their relationships as nodes and edges, respectively, and then obtained a graph. Sun et al. [20] constructed a bipartite graph that regarded entities and entity pairs as nodes, and the edges in the graph existed between entities and entity pairs.

However, these methods have the following problems: First, when building a graph with the help of a dependency tree, the process of generating the dependency tree is error prone. Second, the global and local semantic structure information is not comprehensively considered and then integrated into the model. Therefore, in this study, neither external tools were used nor were entities simply used as nodes. Rather, nodes were constructed for each word in the sentence and then the final graph was obtained. Furthermore, a fusion module was designed to aggregate global and local semantic structure information such that the model can learn more comprehensive semantic structure information. Since the graph neural network acts on the whole graph, this module can better obtain the global information on the sentence compared with the previous research work.

3 Model

Figure 1 illustrates the proposed model, which is composed of three main components: the input representation layer, fusion module, and classification module. Next, all components are described in detail.

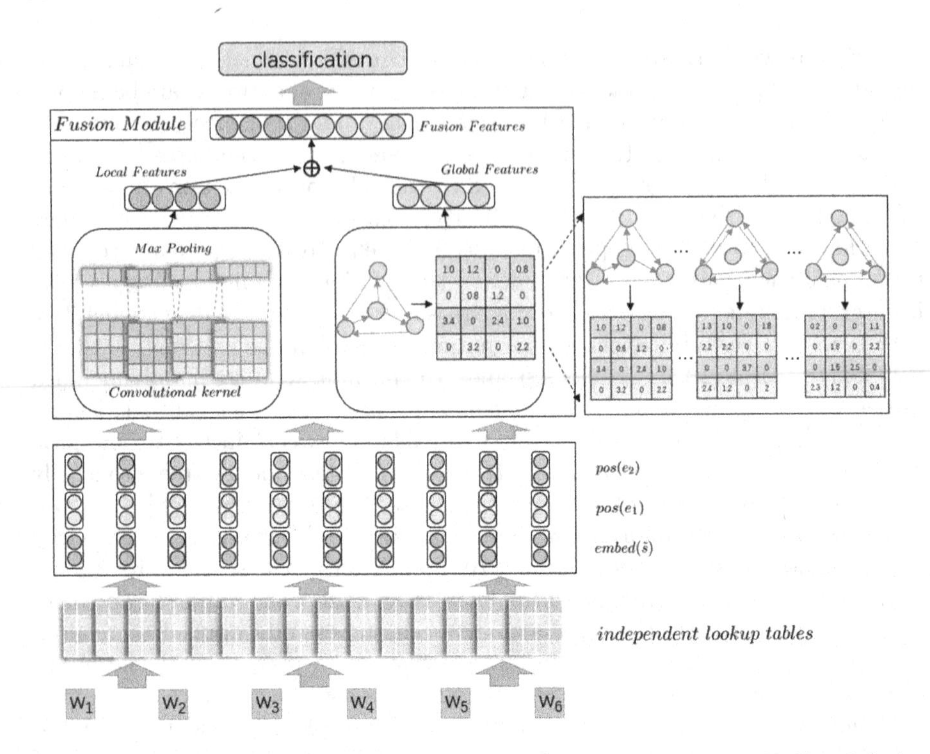

Fig. 1. Overview of the model. The graphs in the model have self-loops by default, hence their corresponding edges are not drawn.

3.1 Input Representation Layer

To highlight the entity, setting labels on both sides of the entities has been used by an increasing number of relation extraction models [21,22], and has also achieved better performance in relation extraction tasks. This study also placed the entity type as a tag on both sides of the entity. Formally, given a sentence, $s = [w_1, ., e_1, ., e_2, .w_n]$ (w_i is a word and n is the length of the sentence) and two entities e_1 and e_2, let *type_of*(e_j) represent type e_j. Then, for the two entities, four tags are obtained. There are <*L_type_of*(e_1)*_1*>, </*R_type_of*(e_1)*_1*>, <*L_type_of*(e_2)*_2*>, </*R_type_of*(e_2)*_2*>. In addition, let T_1, T_2, T_3, and T_4 represent the four labels. Sentence s is then expressed as $\tilde{s}$:

$$\tilde{s} = [., w_i, T_1, e_1, T_2, w_{i+k}, ., w_t, T_3, e_2, T_4, w_{t+m}, .]$$

For the sentence $\tilde{s}$, it is structuralized by e_1 and e_2, which precisely segment $\tilde{s}$ into several parts. Thus, $\tilde{s}$ can be divided into five parts according to the two entities, which are expressed as $left\_of(e_1)$, E_1, $middle\_of(e_1, e_2)$, E_2 and $right\_of(e_2)$.

$$\begin{aligned} left\_of(e_1) &= [..., w_i] \\ E_1 &= [T_1, e_1, T_2] \\ middle\_of(e_1, e_2) &= [w_{i+k}, ..., w_t] \\ E_2 &= [T_3, e_2, T_4] \\ right\_of(e_2) &= [w_{t+m}, ...] \end{aligned}$$

These five parts were obtained for each instance and are referred to as five channels [14]. It was embedded in different lookup tables.

After the sentence is divided into five parts, to obtain the final input representation ($\mathbf{X}$), first, the following vector representation must be obtained:

Word Embeddings. For each part above, a random lookup table was used for embedding. In this manner, each segment will have a corresponding embedding. It is then spliced together in a split manner. Finally, $embed(\tilde{s})$ can be obtained.

Position Embeddings. After the sentence is divided into five parts, the positional relationship of each word relative to the two entities is calculated. Therefore, there are two corresponding positional relationships for each part. Then, five random look-up tables were used to embed the positional relationship generated by e_1 and e_2. Thus, in this process, ten independent lookup tables were used. The five-segment position vector of each entity was spliced together, similar to word embeddings. Finally, the position embeddings $pos(e_1)$ and $pos(e_2)$ relative to the two entities were obtained.

After obtaining the above embeddings, the final input representations $\mathbf{X}$ can be obtained by concatenating them, as follows:

$$\mathbf{X} = [embed(\tilde{s}); pos(e_1); pos(e_2)] \tag{1}$$

3.2 Fusion Module

3.2.1 Global Features

A graph neural network is aimed at extracting information from an entire graph. It can obtain semantic structure information between nodes from a global perspective. Therefore, in this study, a graph neural network was used to extract the global semantic structure information of the sentences. For sentence $\tilde{s}$ with the entity type tag, a graph was constructed as follows: First, each word was treated as a node. Then, the edges and their weights between nodes were randomly initialized. Then, the graph $G = (V, E)$ was produced, where V is the set of nodes and E is the set of edges. Because the GCN [23] can effectively capture the semantic relationship between nodes, it can update the information of the current node according to the information of neighbor nodes. Following multiple layers of GCN, it can be assumed that the information of each node contains the information of the other nodes in the entire graph.

For an L-layer GCN, let $\mathbf{H}^{(l)}$ be the output representation of the l-th layer. Then, the graph convolution operation process is as follows:

$$\begin{aligned}\mathbf{H}^{(l+1)} &= GCN(\mathbf{A}, \mathbf{H}^{(l)}, \mathbf{W}) \\ &= \sigma(\widetilde{\mathbf{D}}^{-\frac{1}{2}}\widetilde{\mathbf{A}}\widetilde{\mathbf{D}}^{-\frac{1}{2}}\mathbf{H}^{(l)}\mathbf{W}^{(l)})\end{aligned} \tag{2}$$

where $\mathbf{A}$ is the adjacency matrix and $\widetilde{\mathbf{A}} = \mathbf{A} + \mathbf{I}$. Further, $\mathbf{I}$ denotes the identity matrix, $\widetilde{\mathbf{D}}_{ii} = \sum_j \widetilde{\mathbf{A}}_{ij}$ denotes the degree matrix, $\mathbf{H}^{(l)}$ is the node representation of the lth layer, $\mathbf{W}$ is the parameter matrix that participates in training, and σ is the activation function (e.g., ReLU). In addition, $\mathbf{H}^{(0)} = \mathbf{X}$.

During the entire training process, adjacency matrix $\mathbf{A}$ changes dynamically; that is, the adjacency matrix is learned. When constructing the initial graph, in previous studies [11], where if there is an edge between two nodes, then $\mathbf{A}_{ij} = \mathbf{A}_{ji} = 1$ otherwise, $\mathbf{A}_{ij} = \mathbf{A}_{ji} = 0$. In contrast, here the edges and their weights between nodes were randomly constructed. Thereafter, the weights were changed during model training. For each sentence, each word was treated as a node. Then, the edges between nodes are equivalent to the relationship between words, which is not easy to obtain at the time of the initial construction of the graph. Moreover, in the process of model training, if there is a strong relationship between two words, the weight of the edge between them increases; otherwise, it decreases. Thus, the proposed model can automatically learn the relationships between the words.

Subsequently, the global features $\mathbf{h}_{global}$ can be obtained after the multilayer GCN.

$$\mathbf{h}_{global} = \mathbf{H}^{(L)} \tag{3}$$

3.2.2 Local Features

Although graph neural networks offer the advantage of obtaining global semantic structure information between sentences they are not good at obtaining local semantic structure information in sentences. However, the local semantic structure information in the sentence can be captured well by the fixed-size convolution kernel in the convolutional neural network. Therefore, to obtain more comprehensive semantic structure information in the sentence, convolutional neural networks can be used to obtain the local semantic structure information of the sentence. Specifically, for a given $\mathbf{X} = [X_1, ..., X_m]$ (m is the total length of the five channels.) and filter $\mathbf{W}_f$, the convolution operation is as follows.

$$h_c^f = \sigma_f(\mathbf{W}_f[X_c, ..., X_{c+k}] + b) \tag{4}$$

where σ_f denotes a nonlinear function and b is a bias term.

Then, for $\mathbf{X}$, the convolution operation can be expressed as

$$[h_1^f, ..., h_{m-k+1}^f] = Conv(\mathbf{X}) \tag{5}$$

where k denotes the number of convolution kernels for a specific $\mathbf{W}_f$. This was followed by a pooling operation.

$$\mathbf{h}^f = Pooling([h_1^f, ..., h_{m-k+1}^f]) \tag{6}$$

Table 1. Details of datasets used. Please see Sect. 4.1 for more details.

Split	ACE05	CLTC
train	83293	13462
val	13779	1347
test	13780	1675

For the $\mathbf{h}^f$ obtained with different convolution kernel sizes, they were spliced to obtain the local features ($\mathbf{h}_{local}$) through the convolutional neural network:

$$\mathbf{h}_{local} = \bigoplus_{i=1}^{r} (\mathbf{h}_i^f) \tag{7}$$

where r denotes the number of convolution kernels.

3.2.3 Fusion Features

Given global features ($\mathbf{h}_{global}$) and local features ($\mathbf{h}_{local}$), the purpose of relation extraction is to predict the type of relationship between entities through these two features. Therefore, to better express the information of these two features, $\mathbf{h}_{global}$ and $\mathbf{h}_{local}$ were concatenated to obtain fusion features ($\mathbf{h}_{fusion}$) for classification.

$$\mathbf{h}_{fusion} = [\mathbf{h}_{local}; \mathbf{h}_{global}] \tag{8}$$

3.3 Classification Module

After obtaining the fusion features ($\mathbf{h}_{fusion}$), $\mathbf{h}_{fusion}$ was fed to a fully connected layer, followed by a softmax layer, to calculate the distribution $p(y\|\mathbf{h}_{fusion})$ of all relation categories.

$$p(y_c\|\mathbf{h}_{fusion}) = softmax(\mathbf{W_c}\mathbf{h}_{fusion} + \mathbf{b_c}) \tag{9}$$

where $\mathbf{W_c}$ is the parameter matrix of the fully connected layer and $\mathbf{b_c}$ is the bias term of a specific category. Finally, the classification result was the relationship category with the highest probability.

The loss function ζ was designed to minimize the cross-entropy value as follows:

$$\zeta(\theta) = -\mathbf{y}log(\widetilde{\mathbf{y}}) - (1-\mathbf{y})log(1-\widetilde{\mathbf{y}}) \tag{10}$$

where $\mathbf{y}$ is the true category probability distribution and θ is the parameter to be learned in the entire neural network.

4 Experiments

4.1 Datasets

In the experiment, two datasets: ACE2005 English (ACE05) and Chinese Literature Text Corpus (CLTC) were used. The data analysis for each dataset is

Table 2. Comparison results with other models on the ACE05 dataset.

Model	P (%)	R (%)	F1 (%)
K [25]	63.50	45.20	52.80
G&J [26]	77.20	60.70	68.00
FCM [27]	71.52	49.32	58.26
M&B [7]	70.10	61.20	65.30
C&M [28]	69.70	59.50	64.20
DRPC [29]	72.10	63.49	67.52
GCN (D+H) [20]	68.7	65.4	67.00
W&L [30]	–	–	70.4
BERT-Z&H [21]	–	–	73.10
Our	84.71	71.52	**77.56**
Our_BERT	87.09	79.79	**83.28**

summarized in Table 1. Below, a detailed introduction to each dataset is provided.

ACE05. The ACE2005 English dataset is a standard dataset used for relation extraction, which was obtained from broadcasts, news, and web logs. It contains seven entity types and six relationship types, and each entity has subtypes and head tags in addition to the type. The English dataset contains 506 documents and 6583 positive data.

CLTC. The Chinese Literature Text Corpus dataset [24] was collected from Chinese literature. The seven entity types and nine relationship types contained in it were manually annotated according to the gold standard. The corpus contains 13462, 1347, and 1675 data values for training, validation, and testing, respectively.

4.2 Hyper-parameter Setting

The hyperparameters were manually tuned for the validation set. The dimensions of the word embedding and position embedding were 300 and 25, respectively. For the number of layers of GCN, we follow [18] to set as one layer. For the length of each segment, except E_1 and E_2, experiments with different values were conducted (10, 28, 78, 128, 178, and 228). The final ACE05 and CLTC datasets were chosen to have lengths of 28 and 128, respectively. Further, the lengths of E_1 and E_2 are set to four each, while the size of the convolution kernel in the CNN was 7, and the number of convolution kernels was 60. Subsequently, parameter optimization was performed using Adadelta with original learning rate of 1.0 and a decay rate of 0.9. In addition, L2 regularization was used with a parameter of 1e−5 to avoid overfitting.

Table 3. Comparison results with other models on the CLTC dataset.

Model	F1 (%)
SVM [32]	48.90
RNN [33]	49.10
CNN [16]	52.40
CR-CNN [34]	54.10
SDP-LSTM [6]	55.30
DepNN [35]	55.20
BRCNN [36]	55.60
SR-BRCNN [37]	65.90
Our	**69.04**
Our_BERT	**73.41**

4.3 Overall Performance

Tables 2 and 3 show the results of the proposed model for the two datasets. In addition, the results of "`Our_BERT`" after BERT [31] pre-training are presented in the table.

Model Evaluation on ACE05. Table 2 presents a comparison between the proposed model and other benchmark models on the ACE05 dataset. Among them, DRPC [29] indirectly used a dependency tree to predict the dependency relationship between words and the relationship between entities, thereby helping to capture text information outside of syntactic information and generalization of cross-domains. GCN (D+H) [20] is also a model constructed using GCN; however, it regards entities and entity pairs as nodes and constructs a bipartite graph. In contrast, in this study, different methods of constructing graphs, which are better than either the graph neural network model or the previous neural network model, were used. Therefore, by using the information of the sentence to construct the graph, the model can converge better, thereby improving the final performance.

Model Evaluation on CLTC. Table 3 presents a comparison between the proposed model and other benchmark models on the CLTC dataset. Many previous models have used convolutional neural networks and recurrent neural networks on this dataset. To reflect the generalization ability of the proposed model, this study experimented with the proposed model based on a graph convolutional neural network on this dataset. As evident from the table above, the proposed model exhibited better performance than other benchmark models. Thus, the proposed model exhibited good performance on both the English and Chinese datasets.

Therefore, it can be observed from this table that the proposed model outperformed all the above results. This performance can be categorized into the

Table 4. Results of ablation experiments. The length of each part of the ACE05 and CLTC datasets are 28 and 128, respectively.

Model	ACE05			CLTC		
	P (%)	R (%)	F1 (%)	P (%)	R (%)	F1 (%)
Our	84.71	71.52	**77.56**	70.81	67.36	**69.04**
- pos	82.89	68.26	74.87	67.35	69.06	68.19
- global	85.30	67.74	75.51	69.49	65.77	67.58
- local	81.97	70.21	75.64	65.64	62.94	64.26
- local&pos	80.07	68.98	74.11	62.82	61.05	61.92

following aspects: First, the model constructed a graph for each sentence, considering each word in the sentence as a node, exploiting the information of the sentence itself. Second, in terms of model input, a sentence was divided into multiple channels according to two entities and tag information was added, which effectively captured the structure information of the sentence and strengthened the model's recognition of the entity's location. Third, the fusion module in the model captured global and local features, thereby obtaining more comprehensive semantic structure information for sentences. Therefore, this comprehensive semantic structure information enhanced the discriminative ability of the proposed model.

4.4 Ablation Study

Further, an ablation study on two datasets was conducted to demonstrate the effectiveness of each component. The results are shown in Table 4. The proposed model mainly includes position information, global features, and local features. Therefore, in this ablation experiment, these modules were subtracted. "-pos," "-global," "-local," and "-local&pos" represent the removal of position information, global features, local features, local features and position information, respectively. Consequently, it can be observed from this table that regardless of which part of the model is removed, the final result is lower. This shows that every part of the model is essential.

4.5 Effect of Length of Each Part

In this section, the influence of the lengths of $left\_of(e_1)$, $middle\_of(e_1, e_2)$ and $right\_of(e_2)$ on the entire model were analyzed. The lengths of E_1 and E_2 are explained in Sect. 4.2 and were set to four uniformly. The lengths of $left\_of(e_1)$, $middle\_of(e_1, e_2)$ and $right\_of(e_2)$ differ for different sentences. However, for simplicity, these were set to the same length. A detailed experiment was conducted for a specific length in this section. The results are presented in Fig. 2.

For the ACE05 dataset, experiments were conducted with lengths of 10, 28, 78, 128, 178, and 228. The results are shown in Fig. 2a. It is evident that a length

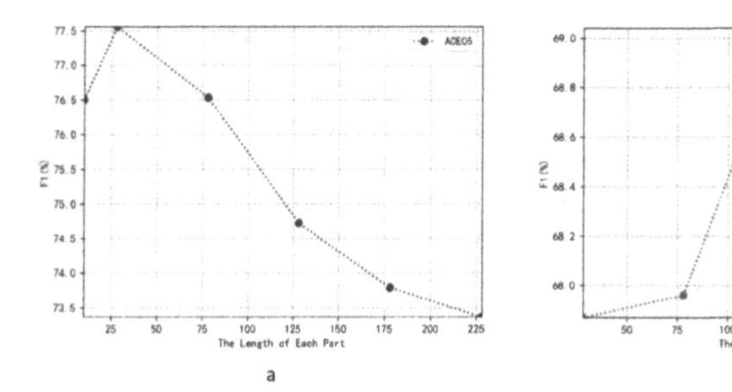

Fig. 2. The results of different length of each part on ACE05 and CLTC dataset.

of 28 can obtain the best result on this dataset. After analyzing the data, it was found that the maximum lengths of the training, validation, and test sets were 97, 91, and 87, respectively, in the three segments. Therefore, for this dataset, the length of each of the three segments should not be particularly long; thus, a length of 28 can make the model converge better.

Figure 2**b** presents the analysis results for the CLTC dataset. For this dataset, the effects of lengths 28, 78, 128, 178, and 228 on the entire model were analyzed. A clear peak was observed at 128. The model converged the best at this length. Similar to the ACE05 dataset, the data of this dataset was also analyzed and it was found that the maximum lengths of the training, validation, and test sets in this dataset were 1874, 765, and 519, respectively. In addition, the sentence length of this dataset was slightly longer than that of the ACE05 dataset. Therefore, its best length was longer than that of the ACE05 dataset.

5 Conclusion and Future Work

This study proposed a L-GCN model that regarded each word in a sentence as a node and dynamically learned the relationship between nodes. Furthermore, a fusion module was used to aggregate global semantic structure information and local semantic structure information such that the model learned more comprehensive semantic structure information. Further, experiments on ACE05 and CLTC datasets verified the effectiveness of the proposed model. In the future, ways to use the information of the sentence to construct novel graph structures to improve the performance of relation extraction tasks will be explored.

References

1. Wang, L., Cardie, C.: Focused meeting summarization via unsupervised relation extraction. arXiv preprint arXiv:1606.07849 (2016)
2. Distiawan, B., Weikum, G., Qi, J., Zhang, R.: Neural relation extraction for knowledge base enrichment. In: Proceedings of the 57th Annual Meeting of the Association for Computational Linguistics, pp. 229–240 (2019)

3. Xu, K., Reddy, S., Feng, Y., Huang, S., Zhao, D.: Question answering on freebase via relation extraction and textual evidence. arXiv preprint arXiv:1603.00957 (2016)
4. Xu, K., Feng, Y., Huang, S., Zhao, D.: Semantic relation classification via convolutional neural networks with simple negative sampling. Comput. Sci. **71**, 941–9 (2015)
5. Zeng, D., Liu, K., Chen, Y., Zhao, J.: Distant supervision for relation extraction via piecewise convolutional neural networks. In: Conference on Empirical Methods in Natural Language Processing (2015)
6. Yan, X., Mou, L., Li, G., Chen, Y., Jin, Z.: Classifying relations via long short term memory networks along shortest dependency paths. In: The 2015 Conference on Empirical Methods in Natural Language Processing (EMNLP) (2015)
7. Miwa, M., Bansal, M.: End-to-end relation extraction using LSTMs on sequences and tree structures. arXiv preprint arXiv:1601.00770 (2016)
8. Veyseh, A.P.B., Dernoncourt, F., Dou, D., Nguyen, T.H.: Exploiting the syntax-model consistency for neural relation extraction. In: Proceedings of the 58th Annual Meeting of the Association for Computational Linguistics (2020)
9. Fu, T.J., Ma, W.Y.: GraphRel: modeling text as relational graphs for joint entity and relation extraction. In: ACL (2019)
10. Vashishth, S., Joshi, R., Prayaga, S.S., Bhattacharyya, C., Talukdar, P.: RESIDE: improving distantly-supervised neural relation extraction using side information. In: Proceedings of the 2018 Conference on Empirical Methods in Natural Language Processing (2018)
11. Zhang, Y., Qi, P., Manning, C.D.: Graph convolution over pruned dependency trees improves relation extraction. arXiv preprint arXiv:1809.10185 (2018)
12. Guo, Z., Zhang, Y., Lu, W.: Attention guided graph convolutional networks for relation extraction. CoRR abs/1906.07510 (2019)
13. Sun, K., Zhang, R., Mao, Y., Mensah, S., Liu, X.: Relation extraction with convolutional network over learnable syntax-transport graph. In: Proceedings of the AAAI Conference on Artificial Intelligence, vol. 34, pp. 8928–8935 (2020)
14. Chen, Y., Wang, K., Yang, W., Qing, Y., Huang, R., Chen, P.: A multi-channel deep neural network for relation extraction. IEEE Access **8**, 13195–13203 (2020)
15. Rink, B., Harabagiu, S.: UTD: classifying semantic relations by combining lexical and semantic resources. In: Proceedings of the 5th International Workshop on Semantic Evaluation, pp. 256–259 (2010)
16. Zeng, D., Liu, K., Lai, S., Zhou, G., Zhao, J.: Relation classification via convolutional deep neural network. In: Proceedings of COLING 2014, the 25th International Conference on Computational Linguistics: Technical Papers, Dublin City University and Association for Computational Linguistics, Dublin, Ireland, pp. 2335–2344 (2014). https://www.aclweb.org/anthology/C14-1220
17. Alt, C., Gabryszak, A., Hennig, L.: Probing linguistic features of sentence-level representations in neural relation extraction. In: Proceedings of the 58th Annual Meeting of the Association for Computational Linguistics, Association for Computational Linguistics, pp. 1534–1545 (2020). https://www.aclweb.org/anthology/2020.acl-main.140
18. Xu, J., Chen, Y., Qin, Y., Huang, R., Zheng, Q.: A feature combination-based graph convolutional neural network model for relation extraction. Symmetry **13**, 1458 (2021)
19. Vashishth, S., Sanyal, S., Nitin, V., Talukdar, P.P.: Composition-based multi-relational graph convolutional networks. CoRR abs/1911.03082 (2019)

20. Sun, C., Gong, Y., Wu, Y., Gong, M., Duan, N.: Joint type inference on entities and relations via graph convolutional networks. In: Proceedings of the 57th Annual Meeting of the Association for Computational Linguistics (2019)
21. Zhong, Z., Chen, D.: A frustratingly easy approach for joint entity and relation extraction. CoRR abs/2010.12812 (2020)
22. Qin, Y., et al.: Entity relation extraction based on entity indicators. Symmetry **13**, 539 (2021)
23. Kipf, T.N., Welling, M.: Semi-supervised classification with graph convolutional networks. arXiv preprint arXiv:1609.02907 (2016)
24. Xu, J., Wen, J., Sun, X., Su, Q.: A discourse-level named entity recognition and relation extraction dataset for Chinese literature text. arXiv:1711.07010 (2019)
25. Kambhatla, N.: Combining lexical, syntactic, and semantic features with maximum entropy models for extracting relations. In: Proceedings of the ACL 2004 on Interactive Poster and Demonstration Sessions, ACLdemo 2004, p. 22-es. Association for Computational Linguistics, USA (2004). https://doi.org/10.3115/1219044.1219066
26. Zhou, G., Su, J., Zhang, J., Zhang, M.: Exploring various knowledge in relation extraction. In: ACL 2005, 43rd Annual Meeting of the Association for Computational Linguistics, Proceedings of the Conference, 25–30 June 2005. University of Michigan, USA (2005)
27. Gormley, M.R., Yu, M., Dredze, M.: Improved relation extraction with feature-rich compositional embedding models. CoRR abs/1505.02419 (2015)
28. Christopoulou, F., Miwa, M., Ananiadou, S.: A walk-based model on entity graphs for relation extraction. In: The 56th Annual Meeting of the Association for Computational Linguistics (2018)
29. Veyseh, A.P.B., Nguyen, T.H., Dou, D.: Improving cross-domain performance for relation extraction via dependency prediction and information flow control. arXiv preprint arXiv:1907.03230 (2019)
30. Wang, J., Lu, W.: Two are better than one: joint entity and relation extraction with table-sequence encoders. In: Proceedings of the 2020 Conference on Empirical Methods in Natural Language Processing (EMNLP), pp. 1706–1721. Association for Computational Linguistics (2020). https://www.aclweb.org/anthology/2020.emnlp-main.133
31. Devlin, J., Chang, M.-W., Lee, K., Toutanova, K.: BERT: pre-training of deep bidirectional transformers for language understanding. arXiv preprint arXiv:1810.04805 (2018)
32. Hendrickx, I., et al.: SemEval-2010 task 8: multi-way classification of semantic relations between pairs of nominals. In: Proceedings of the 5th International Workshop on Semantic Evaluation, SemEval 2010, pp. 33–38. Association for Computational Linguistics, USA (2010)
33. Socher, R., Pennington, J., Huang, E.H., Ng, A.Y., Manning, C.D.: Semi-supervised recursive autoencoders for predicting sentiment distributions. In: Proceedings of the 2011 Conference on Empirical Methods in Natural Language Processing, pp. 151–161. Association for Computational Linguistics, Edinburgh (2011). https://www.aclweb.org/anthology/D11-1014
34. dos Santos, C., Xiang, B., Zhou, B.: Classifying relations by ranking with convolutional neural networks. In: Proceedings of the 53rd Annual Meeting of the Association for Computational Linguistics and the 7th International Joint Conference on Natural Language Processing (Volume 1: Long Papers), pp. 626–634. Association for Computational Linguistics, Beijing (2015). https://www.aclweb.org/anthology/P15-1061

35. Liu, Y., Wei, F., Li, S., Ji, H., Zhou, M., Wang, H.: A dependency-based neural network for relation classification. arXiv e-prints arXiv:1507.04646 (2015)
36. Cai, R., Zhang, X., Wang, H.: Bidirectional recurrent convolutional neural network for relation classification. In: Proceedings of the 54th Annual Meeting of the Association for Computational Linguistics (Volume 1: Long Papers), pp. 756–765. Association for Computational Linguistics, Berlin (2016). https://www.aclweb.org/anthology/P16-1072
37. Wen, J., Sun, X., Ren, X., Su, Q.: Structure regularized neural network for entity relation classification for chinese literature text. arXiv:1803.05662 (2018)

Author Index

Printed in the United States
by Baker & Taylor Publisher Services